DEAD MOOSE ON ISLE ROYALE

DEAD MOOSE ON ISLE ROYALE

OFF TRAIL WITH THE CITIZEN SCIENTISTS OF THE WOLF-MOOSE PROJECT

JEFFREY M. HOLDEN

MICHIGAN STATE UNIVERSITY PRESS | *East Lansing*

Michigan State University Press
East Lansing, Michigan 48823-5245

Library of Congress Cataloging-in-Publication Data
Names: Holden, Jeffrey M. author
Title: Dead moose on Isle Royale : off trail with the citizen scientist on the Wolf-Moose Project / Jeffrey M. Holden.
Description: East Lansing : Michigan State University Press, [2025] | Includes bibliographical references and index.
Identifiers: LCCN 2024057129 | ISBN 9781611865394 paperback | ISBN 9781609177867 pdf | ISBN 9781628955491 epub
Subjects: LCSH: Wolf-Moose Project (Mich.) | Mammal populations—Research—Michigan—Isle Royale | Moose populations—Estimates—Michigan—Isle Royale | Wolf populations—Estimates—Michigan—Isle Royale | Moose—Counting—Michigan—Isle Royale | Wolves—Counting—Michigan—Isle Royale | Moose—Effect of predation on—Michigan—Isle Royale | Wolves—Food—Michigan—Isle Royale | Predation (Biology)—Michigan—Isle Royale
Classification: LCC QL708.6 .H65 2025 | DDC 599.17/8809774997—dc23/eng/20250206
LC record available at https://lccn.loc.gov/2024057129

Cover design by Erin Kirk
Cover photo is Shed Moose Antlers, Isle Royale, by Amanda Griggs.

Visit Michigan State University Press at *www.msupress.org*

Eighty percent of wildlife studies [last for] five years or less; few last more than a couple of decades. The Isle Royale Wolf-Moose Study is pushing seven decades and its contribution to understanding nature and predator-prey interactions is unrivaled. It has also utilized hundreds of enthusiastic and capable volunteers—people who want to make a difference. Together these things have made this study perhaps one of the most important in the world.

—Douglas W. Smith, project leader of the Gray Wolf Restoration Project in Yellowstone National Park from 1997 until his retirement in 2022

CONTENTS

CHAPTER 1

LOOKING FOR DEAD MOOSE FOR FUN

I AM AT THE POINT IN MY LIFE WHERE I AM GETTING PRETTY DARN GOOD AT finding dead moose. So good, in fact, that I list this skill on my resume, right next to where I identify myself as an off-trail backcountry guide. I doubt this has gotten me any jobs, but it is a good ice breaker.

Where do I find dead moose? Well, once or twice a year I visit Isle Royale National Park and lead groups of four or five volunteers, citizen scientists, for a week at a time, into the wilderness looking for them in support of the Wolf-Moose Project. As the name implies, the project tracks the comings and goings of the wolves and moose on the island.

Isle Royale National Park is a large island—roughly forty-five miles long and eight miles wide—in the middle of Lake Superior. It is closer to Canada than it is to the United States. Benjamin Franklin, when negotiating the end of the American Revolution, made sure that the map used to draw the boundaries for our new country had Isle Royale located much closer to the Upper Peninsula of Michigan than it actually is. Therefore, the United States has a large rocky island in western Lake Superior, while the

British—and subsequently Canada—got the smaller Michipicoten Island which is now an Ontario Provincial Park in eastern Lake Superior.

Isle Royale is nominally in my home state of Michigan. It takes some doing to get there. From southeast Michigan, where I live, it is over a ten-hour drive to get to either of two small Upper Peninsula towns that have boat service to the island. That boat trip takes another four hours. If I were to head due south, instead, that minimum fourteen-hour travel time would put me just over the Georgia/Florida state line and enjoying the Sunshine State. Or if I headed east, I would make it to Portland, Maine, with time to spare. Or I could head west to Wichita, Kansas, where I would have an hour and a half to spare for lunch after arriving.

Once on Isle Royale, you are basically on your own in a wilderness, with no roads and very few niceties of civilization. Isle Royale is the least-visited National Park in the contiguous United States with only fifteen thousand to twenty thousand visitors annually. This is significantly fewer than the average attendance of a single Major League Baseball game. But don't let those numbers fool you; Isle Royale is well worth the effort it takes to get there.

I go there for fun as well as to find dead moose.

CHAPTER 2

A HISTORY OF THE WOLF-MOOSE PROJECT

THE WOLF-MOOSE PROJECT IS A WELL-KNOWN—TO NATURAL SCIENTISTS, anyway—predator-prey study and the world's longest continuous study of its type. It started in 1958 and has run in every consecutive year since.

Durward Allen of Purdue University started the project when he got National Science Foundation funding for monitoring the wolf and moose populations on Isle Royale. Allen chose L. David Mech, a recently admitted graduate student at Purdue, to establish and run the program. At the time neither Allen nor Mech knew much about wolves or moose. To be fair, in 1958 there weren't many people who knew much about wolves or moose—or any predators for that matter. Some of the prevailing thinking was that the wolves, who were recent arrivals on Isle Royale, would kill and eat all the moose in short order. People thought wolves were relentless killing machines and that the moose were goners. It was quickly apparent to Allen and Mech that that presumption was incorrect. Wolves are selective and, more often than not, unsuccessful hunters.

In 1958, Allen simply told Mech to count wolves and moose and see what he could find out. The study's first several years consisted of Mech hiking Isle Royale alone, collecting a lot of wolf scat, finding the occasional moose carcass, and in the wintertime flying over the island and counting all the wolves and moose he could see.

Initially the researchers thought a "balance" would be achieved, where interactions between species would find an equilibrium and stable point, and that the primary drivers of the dynamic would be the wolves and the moose themselves. The project was initially planned for only a few years due, in part, to this simplistic view of nature—that and funding for any study, even now, rarely extends for more than three or five years.

But year after year the wolf and moose data suggested a balance was *not* forthcoming. There seemed to be other dynamics in play, in addition to the direct interactions between wolves and moose. Over the years, the equation shifted the story of the wolves and moose to become a story about the impact of weather and climate, insects, the availability of certain plants for food, and stray unforeseen events that have major impacts on populations.

The Wolf-Moose Project is a beautiful example of how science progresses, gathers information, learns or draws hypotheses, reflects on what has been learned, considers other research for concepts and information, and continually questions itself. Science is a process. It is a journey of incremental gains requiring patience and a long-term perspective. The Wolf-Moose Project is exactly that. It involves gathering data, considering that data, and then gathering more data and *complex* data to answer, or at least to investigate, questions that were not considered or even on the radar initially.

Along the way, the Wolf-Moose Project has provided insight into wolves and moose, such as:

- Wolves fail to kill moose more often than not.
- An understanding of long-term trends in air pollution from moose teeth.
- Inbreeding and genetics—both moose and wolves, but especially wolves, show inbreeding on the island.
- How ravens and wolves interact.
- Insight into predation such as, could the wolves really kill off all the moose on the island?

The study has come a long way since Durward Allen told David Mech to simply count wolves and moose and see what he could find.

Why Isle Royale?

A distinctive and defining feature of Isle Royale is its size and isolation. It is an island separated from the rest of the world by—at minimum—fifteen miles of open icy Lake Superior waters. The waters only get above forty-five degrees later in the summer months. This isolation fosters conditions favorable for studying nature.

This isolation is valuable because relatively few species are found on Isle Royale; it is a simplified environment. Isle Royale is inhabited by only about one-third of the mammal species found on the nearby mainland. Some notable species missing from Isle Royale are porcupine, coyote, white-tailed deer, black bear, bobcat, fisher, chipmunk, and red-backed vole.

Moose are present, but they have only been present since about 1900; they likely swam to the island from Canada. And the wolves only appeared around 1950; likely having crossed an ice bridge from Canada. This was a time when almost all wolves in the contiguous United States had been driven to extinction, hunted down, and killed, reducing their historic range to a fraction of its original scope.

Most importantly, wolves are the only predator of moose, and moose are nearly the only prey for wolves. Approximately 90 percent of the wolf's diet is moose. The remaining 10 percent is comprised of beaver and snowshoe hare. Moreover, humans do not hunt wolves or moose on the island. Therefore, the wolves and moose of Isle Royale essentially represent a single-prey, single-predator system. This relative simplicity is not typical of terrestrial ecosystems. Because ecologists are interested in how species interact, Isle Royale is uniquely suited for research because the small number of species translates into a simpler ecosystem that is easier to understand—not easy, but easier.

Isle Royale's size is also important. It is a large island of approximately 210 square miles, forty-five miles long and eight miles at its widest. The island is not so small that the wolves and moose don't have room to move

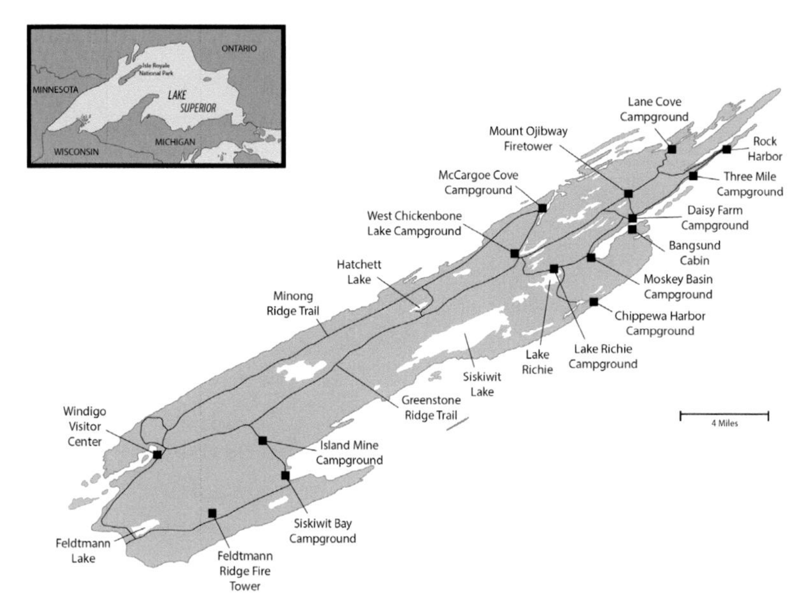

Map of Isle Royale with inset showing location in Lake Superior.

around in, but it is not so large that it cannot be observed and explored effectively by plane during the Winter Study and by foot in warmer weather and the Summer Study.

If you were to design an ideal outdoor laboratory for observing a single predator-single prey dynamic, you couldn't do much better than Isle Royale. There is no other spot on earth quite like it.

Why Citizen Scientists?

While Isle Royale isn't huge, it's not small. There are 210 square miles of essentially total wilderness with trees, rocks, cliffs, swamps, and more, which takes time and effort to explore. To scour the backcountry for dead moose and other information in support of the Wolf-Moose Project is not the work of one or two scientists and a couple graduate students. Monitoring the island requires a significant investment of time—time to hike off trail and stay off trail, and to cover as much ground as possible, year in and year out.

This is where volunteers—citizen scientists—are valuable. Non-scientists can step in and fill a role that scientists do not have the time or resources for. The training and expertise required is fairly straightforward and is augmented by trained group leaders. Groups of volunteers gather data (e.g., data entered on forms, as well as photographs and bones) from the backcountry and bring it to the scientists, who take over.

What makes the Wolf-Moose Project unusual is that citizen scientists volunteer to come to Isle Royale to hike, get dirty, sweat, get bitten by bugs, get rained on, and to trudge through swamps and over ridgelines to look for dead moose. And they enjoy doing it. These citizen scientists, called Moosewatchers by the study, comprise a significant portion of the Summer Study effort. Volunteers have been doing this work for over thirty years, helping the Wolf-Moose Project effectively search the island and telling the story of the wolves and moose of Isle Royale.

CHAPTER 3

HOW I GOT TO BE A GROUP LEADER

THE FIRST TIME I VOLUNTEERED TO LOOK FOR DEAD MOOSE ON ISLE ROYALE was a return to backpacking for me, something I hadn't done in over ten years. There had been that whole "get married, work on the career, ain't got no time to go backpacking" thing going on. Volunteering the first year was a welcome return to packing, but I was tentative, not sure if I'd be in good enough shape to do off-trail hiking. Turns out, being in shape was not really a problem. But did I need new gear? Boy howdy, yes, I needed new gear. And, upon reflection, what was I doing wearing blue jeans while hiking? Nobody would mistake me for a seasoned off-trail moose-finding expert.

It came as a surprise that when that first week of volunteering was over, Rolf Peterson (the study's lead scientist) asked me if I'd consider being a group leader. I'd like to think that my intelligence, calm mature presence, and obvious leadership charisma was what prompted him to ask. Or he figured I was trainable. Or maybe he just asked *everyone* in hopes somebody would say yes and lead.

In reality, I think it was probably none of the above. I suspect that I passed one critical test that first week—and not just once, but twice: I helped butcher two recently dead putrescent moose. Willingly. One moose was liquefied and the other was tick infested and I, alone in our group, helped Phil, our group leader, butcher the moose and get the bones we needed. By "help" I mean I didn't scamper away and I actually handled *stinky* dead moose parts.

Here are my journal notes from that week and the two butcherings:

> Moose #1—Liquefying cow—Amy smells a dead moose. She notices the smell is most pronounced in a little depression in the ground, a sort of mini-sinkhole about ten feet across. I imagine that stink weighs more than typical air and has gathered in the depression. Smelling the dead moose anyplace but the sinkhole is fleeting, illusive. We are day hiking in a relatively easy area: level ground above Chickenbone Lake with only fallen aspen to contend with and long lines of sight.
>
> Amy is at the far left of our picket line and the wind is blowing from her towards the rest of us. Naturally, the bunch of us go into the wind, figuring it is bringing the smell towards us. Our group leader, Phil, being a more knowledgeable, cagey sort, follows the flight of a single raven in the exact opposite direction. Phil puts more stock in the cawing raven than our sense of smell, hikes off, and after one hundred yards walks directly to the dead, stinking moose—and several ravens—on the downward slope toward the Chickenbone Lake.
>
> Even after Phil finds the moose, I remain intrigued by the weird smell phenomena. And this is a moose for smelling. It has been dead perhaps six to eight weeks. A late winter/early spring starvation moose with no dispersion of parts. The moose looks as though it had simply lain down next to a log to rest, like a "meat loafing" cat, and hadn't gotten up.
>
> What remained of the moose was a sort of big moose-shaped bag loaded with putrefying liquefying meat. Much of the fur was gone. Leaving skin that was translucent to almost clear. Ticks on the outside, maggots everywhere, and odd beetles in, on, and swimming in it. Interesting to think the moose was now a sort of gore beetle aquarium.
>
> At first, all of us volunteers hang back in the bright sun, not truly wanting to approach the thing. All week Valerie had been gung-ho on moose stuff, but she has moved away up the hill an additional fifteen or twenty feet; she's not interested in this moose and looks a bit green. Phil has gloves on, the knife out, and various other kill kit items ready to go. Amy announces—emphatically—that she is prepared to take notes. Rich is Rich—he's quietly watching. So I put on gloves to see what I can do to help out with the kill.

It isn't as bad as I had thought it might be. It is in the shade—that's a benefit in the heat of the day—and the smell, although not pleasant, is not crushing. I straddle the back end of the moose and try to figure out if the legs are all there. This was not as easy as it might seem, as the moose had folded itself on top of its legs and then decomposed down and around the legs in an oozy sort of fashion. Rich gets a large log/stick thinking we can lever the moose over to check on the legs. I slide the stick through the rib cage (it just "pops" through the skin) and under the spine. I can then lever and roll the body over and—yup—we confirm all the legs are present and accounted for.

Phil gets the head off in a mildly quick fashion. Then we need a rear leg. These are attached and not ready to come off with a little tug. Or a bigger tug, for that matter. We choose the back right leg, pull it out, and liquid internal organ stuff sloshes into a "sack" of leg skin. Beetles swim back and forth in the goo. I let my brain think this is interesting as I suspect it might otherwise categorize it as good and disgusting and I'd go and join Valerie up the hill a short distance away. The knife is dull and the leg removal takes time. The cuts produce what looks like liquid blood; this surprises me. Eventually the leg comes off and we tromp away from the moose.

We check for ticks. The skinning and bagging of moose parts takes some time. I forget who ends up with the meaty parts, but I didn't want them. I think Rich took them. We hike off.

Back uphill and on top of the flat area we lose the moose smell. . . . Until we get back to the original spot where Amy smelled the moose. I test the depression, and yes, the smell is still there. Amy's spot is a sort of telegraphic smell point.

Moose #2—Tick boy—Our picket line comes out of some woods and we climb a gently sloping ridgeline. In front of us about one hundred feet away is a moose motionless on the ground, except there is a disturbing crawly aspect to it. The moose is in the sun and the day is hot. The moose is covered in ticks. Thousands of ticks give the moose the unsettling odd queasy look.

Amy again announces she'll take notes. Val wanders off. Rich is quiet. So I help Phil as this sort of work is better with more than one person. I ask Phil what I can do to help. He's also hesitant, not wanting to get too close to this many ticks. He suggests that I estimate the number of ticks on the moose. A math problem—I can do that.

I stand over the moose, but not too close, to estimate the number of ticks. I note there are so many ticks that it is almost literally impossible to see the moose itself, except for the nose leather. Oh my. I hold my hands out in a roughly six-by-six-inch "box" over the moose and estimate the number of ticks I see in the "box." I estimate about twenty ticks per "row" and twenty rows in the six-by-six-inch box. That's four hundred ticks. Multiply by four to ramp the number up to ticks per square foot and you have 1,600 ticks per square foot. Yikes.

Then I, Phil, and Amy have a discussion on just how many square feet of surface do you suppose there is on a moose? We decide to first estimated square footage on humans. We discuss this, settling on six-ish feet tall by two to three-ish feet around, plus some [length for] arms, legs, and human crevices. Call it twenty square feet of surface area for a human. Then how much bigger—surface area–wise—is a moose? We settle on four to six times bigger—call it five times bigger—hence about one hundred square feet.

[I] do the math, and it totals 160,000 ticks on the moose. That almost doesn't seem possible. Instead, we simply list that we estimated [more than] one hundred thousand ticks. Yow. No need to wonder how this moose died.

Side note—normally ticks will leave a dead moose in a day or two, so this moose likely died within the last day or two, or perhaps even just a matter of hours prior to us arriving.

I ask Phil if he wants assistance with the butchering. He says no and I don't press the point as I want to get in the shade—I am a bit too hot. Instead I stay close by to help. Phil and I strip down to our boots and pants. No shirt, no hat. Less clothing means fewer spots that ticks could get into.

Phil gets to it. I hand Phil tools. I get a tree branch and attempt to brush ticks away from where Phil wants to cut and sort of help out. Eventually I get into the mix and handle icky-ticky moose parts and pull out the lower back leg—the metatarsus—and put it aside for handling later. Phil does the lion's share of the butchering and, because of the number of ticks, decides to not do as thorough a butchering job as we did with the liquid moose.

Eventually Phil has the head off and we have the moose parts we're interested in. We move away from the moose and the ticks. Phil checks

> me out all over to make sure I don't have any ticks. I do the same for Phil. Then we each head into the woods—in opposite directions—to strip down to nothing. That's what I did, anyway, totally buck naked to *really* check my clothes and body parts. No ticks, thankfully. Phil was also clean.

But did I want to be a group leader? The critical test was the is-he-stupid-enough-to-butcher-a-dead-moose test. I passed that test not once, but twice. But did I want to lead? To learn how to navigate off trail reliably? To be responsible for a group of people I had never met before while we were hours or days away from help if it were needed? To make decisions 24/7 like where were we going to camp for the night? To learn what was required for acquiring moose bones and taking notes? Heck, I wasn't even sure I would come back for year two of volunteering.

I demurred and told Rolf I would think about it, and if I came back the following year, we could talk then. Naturally, I was back for year two. And I said yes.

CHAPTER 4

THE WINTER STUDY

Every year from November 1 until April 15, Isle Royale is closed down. For nearly six months during winter, the animals have the run of the island with no people around, except the several scientists, the ground crew, and the plane's pilot who visit in January and February every year for the Winter Study. Starting around mid-January and continuing for six to eight weeks, the project scientists conduct the winter field studies, *not* volunteers. Several critical activities are conducted during each Winter Study: counting the number of wolves, estimating the number of moose, estimating the rate wolves kill moose, observing behavioral interactions among wolves, and occasionally observing predation events (i.e., wolves killing moose).

Three scientists lead the Wolf-Moose Project: Rolf Peterson, John Vucetich, and Sarah Hoy. Winter Studies in recent years are completed mostly by Rolf Peterson and Sarah Hoy. John is a professor at Michigan Tech and has been with the project since the early 1990s. Sarah is an assistant professor at Michigan Tech, who has been working with the project since 2015.

Rolf has been working on the project since 1970 when he joined the project as a graduate student at Purdue University. In 1975 Rolf became the lead researcher on the project and moved it from Purdue to Michigan Technological University in Houghton, Michigan where it has remained ever since. Rolf is the main scientist the Moosewatch volunteers see while on Isle Royale. For much of the year, Rolf and his wife Carolyn (Candy) live on the island in Bangsund Cabin. They are on the island for around six months every year from April into October. For many volunteers, Rolf and Candy are the face of the project.

The Winter Study's primary method of data collection is through aerial surveys. The project uses small planes, like a Super Cub, with room for a pilot and one observer who rides behind the pilot. The Super Cub is well suited for wildlife surveys because it flies slowly—think 60–70 mph—and can turn tight circles.

Counting wolves via aerial survey involves flying over the island looking for tracks made in the snow by a wolf pack. Once tracks are found, they are followed until the pack is discovered. Once the pack is located, all the wolves are counted several times during each Winter Study to ensure that no wolves are overlooked. There are a lot of photographs of the wolves, which is instrumental in confirming pack size, composition of adults versus pups, the alpha wolves, and so on.

Determining the number of wolves on Isle Royale is fairly easy and there is a high degree of confidence in the number. This confidence is high for several reasons:

- Wolf packs, when resting, are often in the open and it is easy to see every individual in the pack.
- Wolves are territorial with only a little or no overlap between adjacent packs, so if the basic outlines of a pack's territory are known, then a wolf in that territory is pretty certainly not from another pack.
- Last, there aren't that many wolves to count. If, for example, you know there are three packs on the island and you find each pack lounging about and resting, you likely can see, photograph, and count every single wolf on the island.

Counting moose isn't as simple or easy. There is zero chance that aerial fly-overs could count all the moose on the island. There are just too many moose and they all move around and aren't territorial. If you see a

Overhead wolf photo taken from plane during the Winter Study.
SOURCE: WOLF-MOOSE PROJECT 2021–22, ROLF PETERSON

moose one day near Moskey Basin, you might actually see the exact same moose a day later at Lake Richie, but you wouldn't necessarily know it.

Moose "counts" are actually a statistical estimate. The estimates are completed using a sampling of specific census plots on the island. The scientists count the number of moose found in each census plot, and push those counts through a statistical algorithm that generates an estimate and range (e.g., 1,250 moose give or take 250 with a 90 percent confidence interval).

For any given year, the estimated moose counts are just that—*estimates*—but they are probably pretty close. And, because the exact same methodology is used year after year, the *trend* of the counts should be quite accurate.

Aerial surveys also provide an opportunity to calculate how often wolves kill moose, called a Kill Rate. The Kill Rate is in turn is used to generate an annual Predation Rate.

Per Capita Kill Rate (KR) is a calculation of kills per wolf per unit of time. If, during the Winter Study, while flying overhead, the scientists observe that a pack of four wolves killed a moose every thirteen days, the KR would be calculated as:

- Thirty days in a month, divided by
- Thirteen days between observed kills, divided by
- Four wolves

This equals 0.5769 or about 0.6 moose per month per wolf during the winter. The best evidence for Isle Royale suggests that Kill Rates during the summer months are about half of what is observed in the winter. To generate an *annual* Per Capita Kill Rate, it would require the observed rate during Winter Study, along with a factor to take into consideration for the full calendar year.

Predation Rate (PR) is the proportion of the moose population killed in a year by wolves. This rate is calculated over the entire island and the entire moose population and *not* calculated for individual wolf packs. Typical Predation Rates are roughly between 7 and 14 percent. So, if during a single year, the Predation Rate was calculated as 10 percent, that would mean one of every ten moose became a statistic in the Moose Police Gazette on the island.

To close the loop, if wolf abundance (W) and moose abundance (M) are known, then you can calculate the Predation Rate with the formula PR = KRx(W/M), where KRx is the Kill Rate for the entire year.

All of these numbers and rates are foundational and are calculated every year. Knowing the wolf and moose counts along with the Kill Rate and Predation Rate is, to some extent, the backbone of the research. Almost all other data gathered by the project (e.g., temperature, snow depth, tick abundance, availability of forage for the moose) is examined within the context provided by these figures. When joined with predator-prey theory, the Kill Rate and Predation Rate reveal an understanding about how the number of wolves influences the number of moose and vice versa.

All of these numbers are heavily dependent upon the Winter Study—there is no substitute for flying over the island for weeks at a time and direct observation.

Sites where wolves have killed a moose are routinely found during the Winter Study. The location of the dead moose is noted by GPS as the plane flies overhead. And in the summer, volunteers use those GPS coordinates to hike to the dead moose, examine the site, take lots of notes, and retrieve bones. Somewhat surprisingly, the GPS coordinates acquired from an overhead plane are quite accurate and are almost always within about ten to twenty feet of the intended target.

Each year, the Winter Study gathers data and sets the stage for the Summer Study. Where and how many wolf packs are there? Where are the locations of dead moose? Finding these dead moose with known death dates is especially important to the study. Known death dates makes reconstructing the historic moose population much easier. Even though volunteers aren't on the island during the Winter Study, the work done in January and February every year has a lot to say about what volunteers will do in the coming months.

CHAPTER 5

THE SUMMER STUDY

THE PARK OPENS IN APRIL AND DOESN'T CLOSE UNTIL LATE OCTOBER. THIS entire time period is considered part of the Summer Study.

The Summer Study has a number of important goals including collecting moose bones, monitoring vegetation, keeping track of where the island's wolf packs are located, and estimating the numbers of wolf pups born in each pack. Flyovers are *not* part of the Summer Study. Flights over the island during the summer, when the island is open and people are on the island, are limited in order to maintain the wilderness experience. Even if flyovers were permitted, the trees have leafed out and seeing the ground would be difficult.

The Wolf-Moose Project gets help in the summer months from the Park Service as well as vacationers on the island. But the work of collecting moose bones requires off-trail hiking, and lots of it. Looking for moose bone is where teams of volunteers focus their efforts. The Moosewatchers constitute the main effort for collecting as many moose carcasses as can be found. Volunteer and group counts vary from year to year, but most years will see fifty to seventy-five people hiking for weeks at a time covering much of the island in an attempt to find as many dead moose as possible.

> Our valuable Moosewatch volunteers, who put in a week of over-the-top effort to find moose bones, have become an integral part of the research. The specimens they discover are used to track moose mortality and population size, as well as providing an important permanent archive of moose and their environment. For the volunteers . . . it is a challenging and unique adventure in off-the-trail hiking. For the research project, the citizen scientists provide a consistent field effort that would be very hard to maintain any other way.*

Why do the volunteers focus on finding dead moose and not wolves? Moose outnumber wolves on the island. The long-term average is one wolf for roughly every fifty moose. But that ratio is deceptive, as it has varied from one wolf for every sixteen moose in 1980 when there were fifty wolves on the island, to one wolf for every eight hundred moose when the island was down to its last two wolves, due largely to inbreeding, from 2016 to 2018. Because the number of wolves is relatively small and because wolves themselves are not large, the volunteers go into the field specifically looking for dead moose, but not dead wolves. On the rare occasion when volunteers do find a dead wolf, it is a big deal and we do a full backcountry work-up and collect the entire wolf.

How Effective Is the Volunteer Effort at Exploring the Island?

The island is 210 square miles. Off-trail hiking is slow; a pace of 1 mph is typical. Vision while hiking is limited to perhaps ten to twenty feet both left and right—sometimes more, sometimes less—meaning in a perfect world, a single person could visually scour perhaps around four square acres of land per hour. A team of six volunteers in a picket line will cover perhaps 150 to 200 acres every day, or around one thousand acres for a week of off-trail hiking. This is not even 1 percent of the island.

Over the course of the summer, perhaps ten to fifteen volunteer groups will spend a week looking for dead moose. At that 1-percent

* Rolf Peterson, project lead for the Isle Royale Wolf-Moose Project since the 1970s.

rate—optimistically rounding up—roughly 10 percent of the island could theoretically be searched in any given year. This might not sound particularly effective, but fortunately, moose bones hang around and can be found for years after the moose's death. Therefore, over the course of five, six, seven years and longer, a good portion of the island can be searched reasonably well. Additionally, there are portions of the island that aren't thick with moose because there isn't enough food to their liking, meaning the volunteer groups don't search that portion of the island or at least not as often as more moose-rich areas.

While the volunteer effort might look inadequate when viewed as a single team on the island for one week, it is actually quite effective in the long term. Season after season, the volunteer groups cover ground and the miles add up and the finds grow and grow.

When found, most dead moose are scattered bones. Bones that are only slightly scattered suggest a starvation death, while widely scattered bones indicate a wolf kill. Each volunteer group completes a backcountry crime scene investigation (CSI) and collects bones to bring back to the scientists. *Ideally* the skull is collected, as is a metatarsus and bones afflicted with any sort of pathology (e.g., arthritis, breaks, etc.). The metatarsus is the lower rear leg bone in a moose, but in humans is one of the long bones in the foot.

Although wolves kill their share of moose, many moose die of starvation. For instance, the winter/spring of 2021–22 was a year where the starvation deaths were much higher than usual. Most moose starvation deaths occur in the late winter/early spring after a long winter and/or heavy ticks when a moose has exhausted its fat reserves *and* the spring plants have not begun to show. The spring of 2022 was the proverbial perfect storm where it was still snowing on Isle Royale in May and the ticks remained thick into the summer months.

Each season, the Wolf-Moose Project finds and examines roughly seventy carcasses. Volunteers, Park Service staff, and the scientists and students are all part of the effort finding and bringing in bones. Over the years, more than 5,500 moose skulls have been collected. Approximately one-third of all the moose that have been born and died since 1950 on the island have been found and collected. This collection of moose bones and skulls—not surprisingly—is the largest in the world.

This collection of skulls yields a wealth of information. From each set of bones examined, the project estimates the year and sometimes season of death. The age of the moose can be determined by their teeth. The relative health of the moose at death, as well as health in its first year, can be determined from the bones.

Collecting this information from thousands of carcasses enables the Wolf-Moose Project to "reconstruct" the moose population over the years. From that data, estimates of the count, age structure, and relative health of the moose population can be reconstructed and compared from year to year, and further compared and correlated to the wolf population, weather data, presence of ticks, and so on. It is an incredibly rich dataset from which to consider the predator-prey relationship.

CHAPTER 6

THE WOLF-MOOSE PROJECT'S HOME BASE ON ISLE ROYALE

THE WOLF-MOOSE PROJECT'S HOME BASE IS OUT OF AN OLD FISHING COMpound that was built in the 1920s and 1930s on the northeastern side of the island. It is called Bangsund Cabin after the Norwegian fisherman who built up the fishery. Fishing on Isle Royale—along with copper mining and lumber (the island has been stripped of trees before)—is part of the island's history.

Jack Bangsund operated a commercial fishery for several decades and was a squatter; he never held title to any Isle Royale land. Several historic buildings remain from the fishery; there is a main residence, a little cabin, and two smaller sleeping/storage cabins of perhaps eight by twelve feet; these are now named Jack and Howard.

Jack Bangsund died in 1959 and the National Park Service allowed Durward Allen and David Mech, the two scientists who started the Wolf-Moose Project, to use the property as a field work base. It has been the home base for the project ever since. Bangsund Cabin also is an important example of an Isle Royale commercial fishing structure.

The little fishing compound, repurposed for the Wolf-Moose Project, is kind of cute. The numerous small buildings and structures, old ones and newer ones, all seem to have popped up in logical places as if they grew there. There is a kind of organic feel to the place. The main cabin is right

Bangsund Fishery circa 1949. SOURCE: NPS WEBSITE

off of the dock, perhaps fifteen or twenty feet from Lake Superior. It is a two-room cabin of about seven hundred square feet built of hand-hewn logs. The walls and sloping ceiling have pictures, letters, photos, and poems posted all over them. Not surprisingly, most of this material concerns moose, wolves, and Lake Superior. There is a gigantic rhubarb patch immediately in front of the cabin with a flag pole flying a Norwegian flag in honor of Jack Bangsund.

Underneath the cabin is the tiniest of crawl spaces—not big enough for humans, but big enough for critters, usually otters. They are entertaining and love to go from underneath Bangsund Cabin, past the rhubarb patch, and into the water and back.

There is a single rocky trail from Bangsund to the Edisen Fishery. It is a slightly ragged trail along the edge of Moskey Basin. Edisen Fishery is the most intact example of a small family-operated commercial fishery on the island. It is named after Pete Edisen, who lived and worked at the fishery into the 1970s.

If you stay on the trail past Edisen, you will get to the Rock Harbor Lighthouse. The lighthouse was built in the mid-1800s, used for several years, and then de-commissioned because they built another lighthouse positioned

Otter cautiously peeking out from under Bangsund Cabin. SOURCE: JEFFREY MORRISON

more fortuitously for boats on the waters of Lake Superior. The Rock Harbor Lighthouse these days has a slight tilt to the tower and was converted to a museum with displays of the island's history, boat wrecks, and so on. Nobody staffs the museum, so make sure you close the door as you exit.

The rest of the Bangsund Cabin compound consists of structures sprinkled about. There is a new-ish yurt with solar panels, an open-sided boat shelter, a tiny shed just big enough for a circa 1940s cooler (often filled with beer—an important part of the end-of-week celebration for volunteers), and ever-shifting tarps set up between trees for outdoor bucket-and-ladle showers. And of course there is an outhouse. Inside the outhouse is a newspaper article listing numerous euphemisms for outhouses such as dunny, two-holer, long-drop, and so on. All used toilet paper goes in the little waste basket and *not* in the dunny itself. Somebody always goofs this up, and I suspect they have the awful realization about one second too late as their toilet paper drifts down.

Behind Bangsund, to the right of the outhouse, is a series of shelves where antlered moose skulls are displayed, tier upon tier. Not one or two skulls, or even twenty or thirty skulls, but rather—and I am totally guesstimating here—around 150 antlered skulls. It is the largest display in the world of antlered moose skulls. There are regular antlered skulls, along with weird antlered skulls, like the moose with weird squiggly antlers due to goofed-up hormones, the ones with totally asymmetric antlers, and—the ones that fascinate me the most—the super-duper *heavy* antlers, about three or four times heavier than other antlers. Normally antlers will weigh four to six pounds each, and these are vastly heavier. So much so that the volunteer group that found the antlered skull sawed both antlers off so they could carry them more easily, which is not standard operating procedure. Each antler weighs roughly twenty-five pounds. The moose carrying them must have been exhausted when he died.

To the left—as you are on the dock looking at the main Bangsund Cabin—is an open area under a number of pine and spruce trees, where their needles have suppressed the undergrowth, and quite a few tents can

A portion of the moose skulls behind Bangsund Cabin. SOURCE: SUE MORRISON

be set up. This area is used at the end of every Moosewatch week for the volunteers to crash for their last night on the island. It is fun to see over twenty tents all crammed into the area.

Bangsund is where Rolf and Candy Peterson live for the summer season. It has no running water, so water from Moskey Basin is collected, boiled, and filtered as needed. Electricity is carefully generated and maintained through wind turbines, solar panels, battery packs, and meticulous usage monitoring. Rolf and Candy have lived here since the early 1970s and raised their two sons here. They stay on the island from May to October and are at or around Bangsund most of the time. Every year, hundreds to thousands of visitors come to see Edisen Fishery, the lighthouse, *and* the skulls of the Wolf-Moose Project. Candy and Rolf provide tours to the visitors around the Bangsund Compound and talk about the Wolf-Moose Project, the skulls, and how beautiful Isle Royale is. And, if the visitors are lucky, they get some of her fresh-baked orange bread or blueberry muffins.

Last, and most importantly, Bangsund Cabin has chairs. Sure this sounds silly, but after a week of squatting or sitting on the ground or on logs, I absolutely *adore* being able to sit on a chair. Even the picnic table benches are great. It's funny the things you miss when you're in the backcountry.

Other than the few buildings, skulls, and the outhouse, the area around Bangsund Cabin is nothing but forest and Lake Superior—Moskey Basin, really. The only way to get to Bangsund is by boat; *theoretically* you could hike to it, but there are no trails except for the short trails to the Edisen Fishery and the lighthouse, but you can also only get to those by boat too. Any off-trail hike to Bangsund would be miles *and miles* of thick, dense trees and wetland. Nobody hikes into Bangsund unless they are part of the Wolf-Moose Project. I've done it—hiked out of and back into Bangsund—and it isn't fun. It is some of the most difficult hiking on the island.

If volunteers are on the east end of Isle Royale, Bangsund is where Wolf-Moose they start and end their week.

CHAPTER 7

WHAT HAPPENS WHEN VOLUNTEERS FIRST HIT THE ISLAND

VOLUNTEERS GET TO ISLE ROYALE BY TAKING A FOUR-HOUR BOAT RIDE FROM Copper Harbor Michigan. Copper Harbor is *waaayyy* up north at the tip of the Keweenaw Peninsula in Michigan's Upper Peninsula. It is literally the end of the road. Most every volunteer and group leader will be in Copper Harbor the evening before the boat—the *Isle Royale Queen*—goes to the island.

Copper Harbor is a little village with an enclosed harbor, a lighthouse, several hotels, a couple of bars, gift shops, and not much else. The population is supposed to be something like 130 or so, but as you walk around the village, which can be done in about ten minutes, you wonder how it could even be that high. It is small, but it is a fun little town in the middle of beautiful up-north Michigan.

The volunteers meet for dinner at the Mariner North, which has both a hotel and a restaurant. I usually organize the dinner the night before we head to the island, both to meet everyone and to talk to my group and start getting organized. Some of the volunteers and group leaders know each other, but many volunteers do not. We are easy to find as our group of fifteen or so takes up the largest table in the Mariner North's restaurant. Our little group is more than 10 percent the size of Copper Harbor's population.

After dinner in Copper Harbor, I hand out prepared backpacking dinners to everyone. I make it a practice to prepare all backpacking dinners for my group during the week on the island, instead of each person being responsible for their own dinner each night, which always sounds cumbersome and time consuming to me. Each person will carry one dinner for the week and it is easier to hand them out now instead of on the island when things get busy, hectic, and confusing. Also, if I hand them out the night before, I do not have to carry all the dinners to the island—I am always looking for ways to make my life easier.

Generally, the last thing I do before I go to sleep is rearrange my backpack. Even though I've backpacked entirely too much in my life, I still unpack and repack my backpack the night before. *Did I forget some item? Where did I put that such-and-such? Shouldn't this be stashed more conveniently?*

In the morning, the boat leaves at 8 a.m., so everyone is up early to attempt to find someplace to get breakfast. Once on the boat, there is plenty of time to talk to the other volunteers, and if the person wasn't at dinner the evening before, to introduce oneself. A big topic of conversation is almost

Boat ride from Rock Harbor to Bangsund Cabin. Cecilia and Alec are geeked; I am bemused. SOURCE: JACOB DEPPER

always gear talk. "What kind of pack do you have?" "What kind of boots are you wearing?" "What kind of food did you bring?" "Have you been to the island before?" And so on.

For an hour or two there isn't much to see, since we are in the middle of Lake Superior with the closest shorelines ten to twenty miles away. Eventually, we can see Isle Royale across the water, meaning we're still an hour away from landing. And once on the island, we still aren't *there* yet. We usually land at Rock Harbor or Mott Island at the east end of the island. We need a separate, shorter boat ride to get to Bangsund Cabin, several miles down Moskey Basin.

> I was surprised by how remote and isolated [Isle Royale is.] I had backpacked before but this was way more removed. It was an amazing experience feeling so far removed from civilization and feeling like you were a guest in nature.*

At Bangsund, we gear up, augmenting any items we don't have with the project's comprehensive set of equipment. This might include items like stoves, fuel, camp kits, first-aid kits, tents, and so on. I always try to communicate with my group before we head to the island to review what gear everyone has so we can coordinate early on. I like planning, and I try to avoid the confusing hubbub on the island and ensure we have what we need for the week. The communal gear gets distributed to different members of the group so everyone is carrying something, but no single person has too much of the communal gear.

Groups also make sure they have their food organized for the week. Most volunteers bring their own food, so if there is an issue with food on the island, there are limits to what can be done—which is why I prepare dinners and communicate with my group so we don't have these issues. If there are issues, food can be redistributed, and if somebody simply has too much, that can be remedied. But not having enough food is harder to overcome.

---

* Alec Smith, Moosewatch volunteer in 2022 and 2023.

Arriving at Bangsund and unloading gear to start the week. SOURCE: CECILIA VANDEN HEUVEL

So far, the preparation for the week's hike at Bangsund tends to *add* weight to everyone's backpack. I've become quietly militant about backpack weight over time. A volunteer might have too much of their own gear and will say they cannot help with carrying communal gear. Naturally, everyone has to carry some communal gear—I am picky that way. We work through these issues, lightening packs, adding communal gear, and, with luck, have a pack that isn't too heavy, resulting in an easier week of hiking.

Heavy backpacks for the volunteer week aren't unusual, as it is a seven-day hike *and* we have additional gear specifically for the Wolf-Moose Project (e.g., kill kit, emergency radio). A starting weight *under* forty pounds is slightly unusual, and weights *over* fifty pounds are *not* unusual. Bangsund Cabin has a scale to weigh packs and weigh them we do. Ways to skinny down a pack and lighten the load tend to center around removing food (fifteen pounds or more per person for the week is too much), clothing (clothes consume volume in a pack and can be surprisingly heavy), and unnecessary items (e.g., finding out our group has four water filters is probably two filters too many).

The main priority I have as a group leader is to determine our marching orders for the week. Sometimes group leaders get information about where we will be going prior to coming to the island, but usually we get detailed objectives once we are at Bangsund Cabin. I, and the other group leaders, will find out where each group will be during the week; groups do not cover the same areas on the island. We learn where our general area for the week is along with specific sets of coordinates to investigate, usually dead moose coordinates. Often, we have other possible objectives (e.g., investigate old known wolf dens to see if they are active—they are usually not, but it is good to know). Rolf typically gives us more areas and points to investigate than we can do in a week.

Once I know what Rolf expects of our group, I go over my maps to figure out where we might go during the week, what route will be most efficient and wear us out the least, possible campsite locations, and so on. But first I have to figure out where we might be camping that evening. I like to have a notion as to what the weather will be in the coming days, since that can have an impact on any routes I consider. I like the planning aspect, as it is a good puzzle to gnaw on.

Once every group has their gear, food, and objectives for the week, it is almost time to go. Candy takes each group through a checklist of what we should have. Any discrepancies are addressed and fixed. Candy is a good resource for logistics, supplies, and food for the groups.

One set of items I pay attention to is GPS units and compasses. I like to make sure we have at least two sets of GPS units and compasses in the group. If we only had one set and we lost or broke them, navigating by a map alone, while doable, would make for slower hiking and less precise data gathering for the week. The last several years and the advances in smartphones means that GPS and compasses, while extremely useful, can be replaced or at least augmented nicely.

And then we skedaddle, all in about a two-hour chaotic period at Bangsund Cabin. Skedaddling from Bangsund Cabin almost always entails yet another short boat ride, since Bangsund is on an isolated peninsula. Sometimes a group will hike out directly from Bangsund, but most groups have one last boat ride before their week starts.

In 2022, my group consisted of myself and five volunteers. Three of the volunteers were college students studying various flavors of the natural sciences, while the remaining two were a young professional couple with extensive backpacking experience. All were under the age of thirty or thirty-one; I was sixty-two that summer, so it was sort of like hiking with Grandpa.

We left Bangsund Cabin across the water to Daisy Farm and started our hike. Only two of my group had any experience on Isle Royale, and only one had off-trail experience. Basically I had a bunch of young newbies. But these guys all had great attitudes and senses of humor. They were energetic, eager to learn, and everyone was helpful and fun all week long. It was a great week.

Our group had one of the more difficult tasks of the three groups. We had more miles to cover off trail and to more remote and difficult locations. The other groups had fewer off-trail miles and were able to stay in established campgrounds during the week. My group camped off trail all week at multiple different lakes.

We saw a lot of wildlife: live moose, beavers, otters, dozens of pileated woodpeckers, snowshoe hare, numerous bald eagles—seeing them never gets old—loons, mergansers, and fox.

Loons are common on Isle Royale. They tend to avoid humans, especially when nesting, and so off-trail Isle Royale is prime loon territory. The loons love the little lakes that are infrequently visited on the island. Their vocalizations are *fabulous*. They do a high howling that initially might be confused for a wolf, but you quickly learn that loon howls are about an octave higher than a wolf howl. They do a weird tremolo laughing call that sort of sounds like a bad 1950s science-fiction movie special effect. They're great.

Over the six and a half days, we hiked forty-five miles, roughly seven miles a day, with thirty of those miles off trail over ridges, cliffs, and swamps. We saw wildlife every day and covered many of the miles with those forty-five- to fifty-pound backpacks. We found our six dead moose targets. Some of our targets took more time to find than I like, but we found them. We also stumbled upon a number of other "finds," including random moose that nobody knew about until we found them.

Most nights we were serenaded to sleep by multiple loons doing their loon thing.

CHAPTER 8

A STRESSFUL DAY AND NIGHT

Every year, most groups have several people who have never volunteered before look for dead moose. They've never hiked off trail before, and sometimes most or every volunteer in my group has never hiked off trail. Hikers sometimes get lost *on* trail. Getting lost is easy when you are off trail, and it is time consuming and stressful.

When I lead groups, I make sure we stop and take a short break before we head off trail. We get situated, put clothes on or take them off, and get some water and a snack. I do a short primer on what off-trail hiking is like and how we'll spend our week. The main point of the primer is to always know where *I* am, because the group leader—even if lost—is where the group needs to be.

If we are spread out horizontally in a picket line so we cover more ground, thus enabling us to find more dead moose, it is important to always know where the person on either side is, especially the person who is closer to the middle of the line. An attenuated picket line, in thick brush, is an invitation for people to accidentally veer off and get lost. Generally, whoever is navigating is either in the middle of the picket line or, if single file, at the front. As we hike and check the ground for bones, we look up so we don't fall (at least not a lot) and so we know where our team members are. We yell a lot (e.g., the call-and-response *Marco Polo*).

Most people take a day or two to get used to hiking off trail, especially hiking in a picket line. Some people never really get the hang of it, but everyone improves as we hike. Off trail, there are hills, swamps, and downed trees to walk around. When everybody is zigging and zagging, picket lines get attenuated, ragged, and spread out. That is when we yell to ensure everyone is visible to somebody else. If not, we make it so, usually by stopping briefly and having the person furthest from the middle of the picket line pull in toward the middle.

The very first year I led a group, we hiked about an hour on trail before going off trail. In this group, there were two people who knew what off-trail hiking was and three with no off-trail experience. And I, as a newly minted group leader, was hyper-aware that I'd never done the leading thing before. I did have the presence of mind to stop before we left the trail to complete a primer on off-trail hiking and expectations and ask if anyone had any questions.

One question was where we were headed. Off trail, there are, of course, no campgrounds and few "solid" landmarks. My default is to simply pick a point on the map that will get us to where we need to go, eyeballing coordinates from a map using UTMs/WGS84 (a universal metric system that functions similarly to latitude and longitude), and navigate toward that arbitrary point. As I answered the question from my group, I used my finger to point vaguely at the map to a tiny peninsula and shoreline. My finger was big enough to comfortably cover a quarter-mile spread of topography on the map. And off we went, tramping into the woods on the western end of Isle Royale on a chilly, cloudy, rain-threatening sort of day.

Almost immediately, we were too spread out with some people zooming ahead, some being cautiously slow, and some going in the wrong direction. We yelled, we came back together, we got organized, and we went off again. It was like herding cats. We did this twice and after only fifteen minutes, we all yelled came back together, and I said, "Where is Steve?" (not his real name).

Nobody knew, although somebody said when they last saw Steve, he was *way* out in front and moving fast. We stayed where we were, yelling.

Somebody had a whistle and blew it. This is when I learned you cannot hear whistles very far in dense forest. I then ran ahead—as best I could in the woods—and yelled more. I came back to the group.

I decided Steve was ahead of us and our best chance of finding him was to continue hiking, thinking he would be concerned he wasn't with his group and stop somewhere for us to find him as we hiked. We did this in a picket line again, yelling all the way. After another fifteen minutes, we had covered a lot of ground, but we had not found Steve.

What to do? My *first time* leading a group and literally within fifteen minutes of being off trail, I lose a volunteer. This would not look good on the group leader job review.

One of the two volunteers who had previously been on the island and hiked off trail suggested that he split off from our group, head to the shoreline, and then hike the shoreline until he found us, thinking that Steve might have gone to the shore and was hiking there. It wasn't unreasonable. The remaining four of us would continue hiking straight across the land, away from the shore to the spot on the peninsula we were aiming for. The volunteer who suggested this (Clay—who is now a group leader) would have a longer route to get to my arbitrary point, but, as it was on the shoreline, he would eventually find us. In theory.

I didn't know Clay from Adam at that moment, but his plan sounded reasonable. However, I was *super* leery about having another volunteer split off from the group. Clay had a map and a compass, but had never navigated off trail before. We only had my one GPS unit and Clay was not getting that. I eyeballed a bearing (a compass direction) that would take Clay to the shoreline, stressing that when he hit the shoreline, he needed to turn right and follow it until he found us.

And Clay went off. I wondered if I had made the correct decision in letting him go. He seemed competent, but I did not know Clay. The rest of us trudged directly toward my arbitrary spot. It started to rain.

After two hours, it was mid-late afternoon and the four of us made it to the shoreline where I'd been planning on finding a location to camp. The spot we eventually chose was right on the shoreline on a promontory overlooking Lake Superior. As long as Clay was along the shoreline, he should be able to find us easily. It was not, however, exactly at the arbitrary coordinates I had aimed for. The coordinates I'd chosen were *inland* away from the shore about one hundred yards (this is important—remember this). The inland coordinates had been in the midst of a dense stand of cedars. Isle Royale seemed to specialize in short, wide, dense cedars in swamps, which are hard to hike in.

Regardless, we were roughly where we needed to be—and where Clay expected us to be. On the plus side, it had also stopped raining. We, of course, had not found Steve. I was worried about him. Was he horribly lost? Frightened? Hurt? Wondering where we were? Would he set up a tent?

We went about our business, made camp, had dinner, and waited for Clay. Maybe he had found Steve. Another hour passed and it was getting dark and was raining again, Clay finally made it to us in a bedraggled state. He was glad to see us and get dinner. But still no Steve.

I needed to let the project know I'd lost a volunteer. This was 2005—several technological ages ago. The only way to communicate with the project was either in person—not happening—or a line-of-sight radio that weighed several pounds, along with its several-pound spare battery. I never liked those radios. I fired up the hated radio and attempted to get somebody to respond. One minute, two minutes, five minutes. Nothing but static.

My topographic map suggested I wasn't in line of sight because of the big hills between us and the ranger's office. According to the map, it looked like I needed to hike back the way we came, about 1.5 to 2 miles. It was raining harder. I put on rain gear, made sure everyone—who wasn't lost—was in the camp, had eaten dinner, and was ready for a relatively dry evening in their tents. Clay and I were sharing a large two-man tent for the week. Then I hiked off in the rainy night looking for a ridge line back a couple miles. *Sigh.*

By the time I got there, it was 10 p.m. and downpouring in the dark. I was a couple miles from my tent, a volunteer was lost, and it was the first time I'd ever led a group. This was probably the most stressful day of my life; it still holds that special place in my traumatized heart.

I finally got somebody to answer the radio, some lower-level person at the park headquarters who seemed happy to have somebody to talk to. He said everyone was at a party. Why wasn't Rolf answering the radio? Because he was at the party. Eventually somebody who could do something, a park ranger of some sort, got on the radio asked questions including (drumroll) coordinates for where we were camped. *Shit!*

I am an idiot, I thought. I gave them the incorrect, but sort of suggestive of where we were, inland coordinates, letting them know they were suggestive. The park ranger said they would get a boat and come out and see what they could see. It was an opportunity for them to do their search-and-rescue routine. It was early spring and the Park Service had new staff to train in search-and-rescue, so we were simply helping out by presenting a training opportunity, so it was not a complete loss. Yeah, right. Glad to be of service.

At that point, at the top of a ridge, the rain pouring, and a long way from a dry warm tent, I had done what I could. I turned back to our campsite. The *Shit!* moment, while talking on the radio, is where I learned an important lesson. *Always* GPS your campsite the moment you know where you are camping. And if somebody else in your group has a GPS unit, make them do it too. Redundancy is good for critical, foundational things. I knew the basic direction back to my tent, but I did not know exactly where I needed to go, where my group, minus Steve, was.

One step at a time. I hiked toward the suggestive coordinates because they were close to where I wanted to be. I got back to the coordinates in the middle of the dense cedars. It was raining harder. I stopped and hoped I could visually figure out the direction to the shoreline from where I was. Weeping was a considered option. It was late and I wanted to sleep. Instead, I pulled out my compass and, knowing if I simply walked due west, bearing 270, I would hit the shoreline and then could go south, or left, and find our campsite. In retrospect, given that I was about to cry, I am pleased that I figured out this logical and easy solution. I walked due west for several minutes, hit the shoreline, and looking left, could see the promontory and our camp. I was home and only a semi-idiot, but learning.

I made it back to camp and was getting ready to crawl into my tent (move over, snoring Clay) and sleep in a damp fashion when I saw a bright spotlight. Park rangers. It was midnight now, and falling asleep would have to wait. The rangers were in a boat along the shoreline with their super-duper bright spotlight looking for Steve, and approaching our campsite. I went to the water to meet them.

The boat bumped toward the shore. Rolf was on the prow and asked what was going on. I did my authoritative best to sound logical, factual, and like I was in semi-control of the situation. In reality, I was hoping Rolf would tell me to get on the boat and announce that he was taking charge, and I could go home and never return to Isle Royale. That did not happen.

Instead we—Rolf, me, and the park rangers—talked the situation over, like reasonable adults or something. Didn't they know I was panicking? We decided that me and the volunteers would stay in the camp while Rolf and the park rangers would take the boat a bit further up the shoreline and then turn around and head back to camp at Windigo in Washington Harbor. The next day, we would stay in our campsite until we heard from them about if they found Steve. Inaction—yes, I can do that.

The boat pushed off and they headed north to continue their tour up the shoreline. I went to bed. Clay, who never woke up during the ranger boat visit, was totally, deeply asleep and still snoring. I crawled into my sleeping bag. The rain, which had been lessening, decided to do the deluge thing with some lightning thrown in for good measure. This continued for as long as I remained awake. And, while awake, I worried.

Where was Steve? Was he crazy and lost and unprepared for what was occurring? Was he wondering where we were? Wondering why weren't we looking for him? And so on and so on. I was obsessing. It took a while to fall into a fitful sleep.

The next morning was a clear, bright, sunshiny day. If we were looking for a 180-degree contrast to the dark, rainy, stormy night, this was it. We weren't supposed to leave our spot until we heard from the Park Service. The five of us moseyed about, had breakfast, took down our tents, broke down camp, and were ready to go. We sat and waited. And waited.

At noon, the Park Service boat appeared and—holy cats—were we happy to see somebody doing something. The boat bumped up against the shore. Rolf was again on the prow, and so was Steve, looking totally calm like nothing had occurred. Part of me wanted to torch/castigate/destroy him. But then the more mature, calm me figured I ought to find out what happened. Although upon reflection, that didn't really help.

The day before, Steve had seen me point to the spot on the map and figured he would just hike there and we'd find him. *What?* When he said this, I was incredulous—it was an arbitrary spot on the map and Steve—who didn't have a map or compass—just hiked off into the unknown.

To compound the hurt, apparently Steve had been up the shoreline about half a mile from where we were, just beyond where the Park Service searched the night before. He was under a gorgeous stand of pines, next to the shore in a nice, dry spot with moss making a soft bed. He had a fine night.

Steve hopped off the Park Service boat with his gear. He was ready to go. I told him to go to the other volunteers and let them know we'd be leaving shortly. I talked with Rolf and the Park Service.

Apparently, Steve had gotten up early that morning and hiked to the Park Office about six miles away, where we'd started the day before. He had met Rolf and a park ranger or two and said his group had lost him. They wanted to read him the figurative riot act, but they were so glad to see him, they left that for later. They made sure he was fine, had his gear, got him on the Park Service boat, and brought him back to us.

So that was it? He's mine again? We said our good-byes to Rolf and the Park Service people, and we headed north back toward where Steve had spent the previous night. Ironically, one of the kills we needed to find was very close to where Steve had camped.

For the rest of the week, a long week, I watched Steve like a *hawk*. Initially, I told him he couldn't be ahead of our group and he had to be behind at least one person. It very quickly became apparent that was not going to work. Steve was hyper, twitchy, and couldn't be held back. We somehow ended up at an agreement that he would zig-zag in front of us; mostly in front of me. Essentially, he hiked about twice the distance as the rest of us, but he managed to do this and not get lost, which was, of course, the goal.

Coincidentally, on our last full day, within roughly one hundred yards from our campsite, another person in our group got lost. We were all together and approaching our campsite, seeing our tents, and simultaneously found a dead moose. It was the end of a long day of hiking and this person, who will remain unnamed, was tired and asked if they could go back to camp while the rest of us did the dead moose thing. I remember distinctly looking at the tents as I said, "Sure. Go back to camp and stay there, and we'll be there in a bit." Somehow, this person managed to go toward the tents, hang a hard left, hit the shoreline, and kept on going.

We got into camp with moose bones, tired and hungry, and the camp was empty. We did the yelling thing again. For the record, that only occasionally works. After two hours we found this person down the shore roughly half a mile, just sitting by Lake Superior, waiting for us to arrive.

The next day, we hiked back to Windigo where our week had started. On the side, I told Rolf to never let Steve and the other person who got lost ever hike for the Wolf-Moose Project again. They have not. And Rolf informed me that they had talked to Steve's wife while he was lost and she had said something to the effect of we shouldn't worry about him running off; he did that all the time. He would simply race ahead of whomever he was with, and they would not see him for hours, but eventually—inconveniently—he'd come back.

Interestingly, in the subsequent twenty-plus years of leading volunteer groups, I've only ever had two other people get lost while hiking, and they were never lost for more than five minutes. They both knew *immediately* they were lost, and as they were yelling (boy, were they yelling), you could tell they were frightened and concerned.

Being lost is stressful.

CHAPTER 9

WHY FINDING DEAD MOOSE CAN BE TRICKY AND HOW TO SEARCH FOR THEM

AN IMPORTANT PART OF THE WOLF-MOOSE PROJECT IS FINDING AND STUDYING dead moose (otherwise it would just be the *Wolf* Project). The greater the number of dead moose you can find, the more you can learn about the moose population through the years. The problem is that first, you need to find the dead moose. There are four issues: moose avoid people, wolves disperse dead moose, things grow and hide in dead moose, and even dead, decomposed moose are large.

Moose avoid people. When alive, moose do their darnedest to keep away from people as well as wolves. Moose do have a *slight* preference for people over wolves, I imagine, but they definitely are not attracted to us. Moose hang out in swamps and beaver ponds, and they frequent the edges of little out-of-the-way lakes. They hide in tiny spots you would *swear* they couldn't get into. Crawl into the densest thicket of fir trees and you will find moose poops square in the middle of the smallest space that *you* have trouble getting into. The result is that when moose die—no surprise—they invariably do it in an inconvenient place.

Wolves disperse dead moose. Wolves like to eat moose. When wolves eat, they are not neat about it. Yummy parts include the head and legs, which are easily disconnected from the body, almost as if there were perforated lines on them. Once disconnected, these body parts get dragged away—sometimes for hundreds of yards—for convenient individual devouring. Wolves do this because they get jealous of what the other wolves are eating, or because the pecking order demands distance. Unfortunately, those dragged-away parts are exactly the parts the scientists are most interested in.

Things grow and hide in dead moose. Moose die on the ground in the woods. Plants grow where moose die. Most dead moose aren't found quickly; sometimes it will be years before somebody comes across their bones. The accumulation of annual leaf fall, growing moss, ground cover taking hold, trees falling over, and so on means that over time, dead moose get covered up by the forest. When a moose has been dead long enough, they might be covered to such an extent that they are easily missed even if you walk directly on top of them.

Even dead, decomposed moose are large. Should you overcome the first three obstacles and actually find a dead moose, then the task of carrying the parts of interest out of the forest remains. Moose are large animals. Individual bones and antlers might not seem like they weigh a lot, but once you start tossing multiple moose pieces onto a fully loaded backpack, the weight adds up quickly. It is not unusual for volunteers to have heavier packs at the end of the week than when they started.

How exactly do you search for dead moose? Here are some handy tips. Remember, moose tend to die in difficult places. They rarely die in campgrounds or on trails; it happens but not often. Here is how you explore those inconvenient, off-trail backcountry places:

- Get multiple people willing to tramp through the undergrowth someplace where moose are located, like Isle Royale. Have these people do this for many days and as many miles as they can manage.
- Get the volunteers to bushwhack through the wilderness but *not* in a single file. Spread out and space yourselves so there is fifteen to fifty feet between everyone. This type of picket line covers more ground. The spacing will vary depending on the terrain and the denseness of the undergrowth. Note—you'll spend a fair amount of time simply

trying to keep track of each other and to not let anyone get lost. Keeping track of each other tends to slow down forward progress, meaning the average speed is modest.

- Finally, look at the ground.

Easy peasy, right? Keep in mind, off-trail hiking is tougher, or at least clearly different than walking on a nice, flat surface. You are carrying more weight than usual; at a minimum you will have a daypack of probably ten or fifteen pounds. For a full-pack off-trail hike, you will be carrying at least forty to forty-five pounds and as much as sixty-five or seventy pounds; moose bones weigh a fair amount.

Other variables to finding dead moose are seasonal and depend on where you are on the island. Springtime is much better for finding dead moose—once summer comes along, the ground cover grows and thickens and it becomes more difficult to see the ground. Heck, sometimes it is difficult to see your feet as you hike. Different spots on the island tend to have more or fewer dead moose due to forage. For instance, if you are on the western end of the island by Feldtmann Ridge to Lake Halloran, you are almost guaranteed to find double-digit counts of dead moose. Whereas if you go to the middle of the island—where the big burn of 1936 covered about 20 percent of the island—you will have a lovely hike in aspen and birch, but find few, if any, dead moose because the foliage that has grown back since the fire are not preferred moose food.

As you walk, you are not just interested in the spots your feet are tromping on—you need to look left and right anywhere from ten to twenty-five feet in each direction. All the while you have to climb over and under messy tangles of dead trees that have fallen while avoiding the swampy bits. This all makes hiking more difficult and slower. Off-trail hiking of 1 mph is considered "fast" and you will stumble now and then. It is unusual to make it through an entire week without falling; most people fall multiple times. In 2022, I fell four times along with many near-falls where I would catch myself and then claim I did *not* fall. I have rarely had a year where I did not fall.

Remember, this is fun.

CHAPTER 10

LOOK FOR DEAD MOOSE, FIND DEAD WOLF

Every year the Winter Study takes place, and as the scientists fly over the island, they find moose the wolves have killed. Sometimes the scientists—or the ground crew—will hike, snowshoe, or cross-country ski to these dead moose during the Winter Study. They GPS the more remote and/or difficult-to-get-to moose from the plane flying overhead. And come the warmer weather, summer interns and Moosewatch volunteers hike off trail to find them and collect their bones and information.

In 2006, one of these hard-to-get-to dead moose—killed by the Chippewa Harbor wolf pack—was on the north slope of the Greenstone Ridge. The north slope is the steeper, harder-to-climb side of the ridge, where one hundred feet north or south might also be one hundred feet up and down in elevation. My group was assigned to go and get this moose.

At some point during the week, we hiked to the GPS location to search for the dead moose. Rolf Peterson knew exactly where the moose was and described to me the location, as well as the trees and rocks nearby. The island was covered with trees and rocks, and the perspective from the plane overhead differed significantly relative to being on the ground, but it didn't make no never mind. Rolf knew exactly where this moose was, and we needed to get it.

Normally with GPS coordinates, you can walk right up the dead moose and—even with coordinates taken from overhead in a slow-moving airplane—find the moose either instantly or within a minute or two. This moose was an exception. To repeat: steep slopes, with significant elevation change, makes it harder to search for things.

We searched for fifteen minutes, then thirty, forty-five, and sixty minutes. Nothing. Bubkes. Because I knew there was a moose at this location and that the location was difficult to get to I really, really, *really* wanted to find the moose and not disappoint Rolf. Finally, one of my group, Steve (this time it's really his name) said he had found something, but it wasn't a moose. He was about fifty or sixty feet above me and I needed to pick my way up the rocky slope to get to him.

"What is it?" Steve asked when I got there.

"That's a wolf," I said. And he asked if we were interested in it because, of course, we were looking for a dead moose. *Oh yeah*, we're interested. The wolf had been killed by other wolves and was either partially eaten and/or ripped apart a fair amount with its back legs missing. Before moving anything, we took photos, collected data, and then gathered the entire wolf, not just a few bones. We didn't have any bags large enough for the wolf so we used my backpack cover (a waterproof cover to protect a backpack) and bundled the wolf into it. The wolf, once bundled up, didn't make for a very big beastie. And because it was a wolf—a rare find—I wanted it totally inside somebody's backpack and not chancing losing it by having it on the outside of a pack. Thomas in our group had space, so we slipped the wolf inside his pack.

The map suggests a story of what had happened. The Chippewa Harbor Pack (CHP) had territory covering the mid-east of the island, and the East Pack (EP) was to the east. The shaded section was a portion of the island where those two packs were competing for territory. The moose we were after was the square above the P in CHP, just to the west of the contested shaded area. What likely occurred was that an East Pack wolf

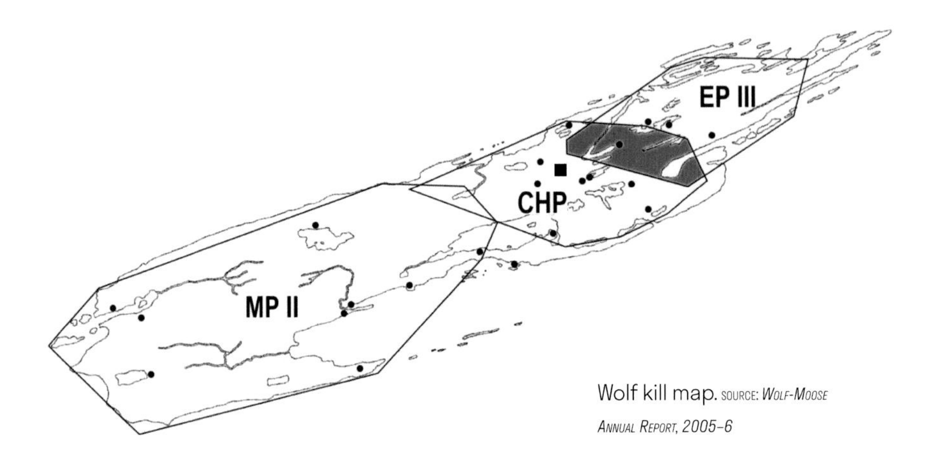

Wolf kill map. SOURCE: *WOLF-MOOSE ANNUAL REPORT, 2005–6*

Rolf holds up the head of the dead wolf. SOURCE: JEFFREY HOLDEN

decided to visit a wee bit outside of his territory, and poach a bit of the moose the Chippewa Harbor Pack had killed. The Chippewa Harbor wolves did not take kindly to this, and killed the poacher.

My group never did find the moose we were looking for, but, since we had found a wolf instead, I knew Rolf would forgive me, especially since it was a wolf we would *not* have found if we'd found the moose quickly. I like to think of it as productive incompetence. A day or so later when we were briefly on trail going to find something else somewhere else, we chatted briefly with a hiker and told him of the wolf we had found.

At the end of our week, Rolf picked us up in a boat at Moskey Basin to take us back to Bangsund Cabin. His first words to us were, "I hear you found a wolf." The backpacker grapevine was efficient. The photo is of Rolf showing the healthy, unbroken teeth of our young wolf—a young wolf who poached one moose too many.

CHAPTER 11

OFF-TRAIL BACKCOUNTRY BACKPACKING AND CAMPING

THE MAIN GOAL OF VOLUNTEERING FOR THE WOLF-MOOSE PROJECT IS TO FIND dead moose. Dead moose die off trail and occasionally on trail, but those are found quickly. Volunteers go to Isle Royale to find dead moose *off trail*, where things aren't as comfy or nice as on trail.

First, here are some things that off-trail backcountry backpacking is *not*. It isn't setting a tent up in the backyard where you can have a hot cocoa from the house when it gets dark or cold. It isn't car camping where you have a cooler (with beer!) and if you've forgotten something, you can hop in the car and drive into town. It isn't parking the car at a trailhead and hiking a well-defined trail to an established campsite with a tent pad, picnic table, firepit, and perhaps an outhouse.

Off-trail backcountry camping is when you are in the wilderness, hiking with a full pack, starting weight of at least forty pounds—probably more like forty-five or fifty pounds, to be realistic. This weight includes all your gear, food for one week, water, and communal gear. Off-trail hiking means regardless of the weather, going through or around everything you meet.

Crossing a beaver dam off trail. SOURCE: ALEC SMITH

Think swamps, rocks, ridgelines, creeks, and beaver ponds. Likely you don't know where your tent will be that evening, or even if your tent will have a dry, flat spot for set up. You hope you have a good source of water nearby, which will still have to be filtered. This can consist of pumping it, squeezing through a filter, and/or using a gravity filter, which is my favorite.

Your group deliberately leaves a perfectly good trail, and you head into bushwhacking country. Off trail, there is no easy hiking, instead you have to crawl under deadfalls, over trees, step into and/or around swamps, with the occasional fall.

There are no nice campsites or outhouses, which are suddenly viewed differently; they've become a "luxury" just by leaving them behind. And there are definitely no other people. In the backcountry, you are on your own and it is unlikely you will see people—other than your group—until you return to trails and get back to campgrounds. Any issues or injuries you have during this time are for you to solve on your own. Help is many hours or possibly even days away.

Off-trail backcountry packing can be confusing and uncertain. Where do you camp and how is that decided? Mostly it is the group leader who decides, unless somebody in the group has experience with this sort of thing. I've backpacked off trail in a number of other parks and know that the "perfect" off-trail spot—while it might theoretically exist—isn't something to aim for because you probably won't find it and you might waste a lot of time looking for it. Take the reasonably good spot and be happy about it. I've seen enough of Isle Royale to know that theoretical "perfect" spot might not even exist on the island. The terrain is rockier and swampier than most places. So when you find a "reasonably good" spot on Isle Royale, grab it.

If it is midday and you stumble onto a sight that has water nearby say within one hundred yards, enough open spots for your tents, and a position to allow you a fairly easily hike to wherever it is you want to hike the next day, that is more than good enough to set up camp. And truth be told, you can always find enough spots for tents; they might be spread out over a bigger area than you'd like, but it can be done.

As an example, my group in June 2022 was mostly full of new volunteers to the project with little or no off-trail experience. On our first afternoon off trail, they seemed mystified when I announced a spot for us to camp.

South shore, Lake Shesheeb campsite. SOURCE: MEGAN HORODKO

Sure, there was a lake right there—Lake Shesheeb, a pretty little lake with, as it turned out, a lot of wildlife. But the spot had downed branches, undergrowth ground cover all over, and the ground didn't look particularly level. It was nothing like a state park where you pull your car into a parking spot next to the firepit and the nice open flat area for tents. The area by Lake Shesheeb was big enough that when we explored a bit, we found quite a few flat spots, and with a few minutes' effort, we moved larger fallen branches and flattened the undergrowth. Our four tents ended up in the same general area spanning about 150 feet strung along the lake shore. *Voila!* Campsite.

Yes, there were still lots of branches and weeds and little trees all around that made it more difficult to walk around the "camp," but it worked fine. My tent was situated along what looked to be an occasionally used moose trail complete with fist-sized rocks that I finagled my tent, and sleeping bag, around—but it was flat. We all found spots, put up a tarp (in case of rain), and explored a bit.

After our tents were set up, a cow moose strolled by, unconcerned, checking us out. Exciting—our first moose for the week. We camped at this location and loved it.

> Moosewatch [is] awesome . . . but probably isn't for everyone . . . let me tell you about it and how [you] could have fun doing it. You will learn so much. It's a fantastic island . . . [But] some people can't even fathom peeing outside . . . you mention tent, leave no trace, etc.—they just look at you like you are crazy. And some people are just totally into whatever I tell them . . . and want to do it.*

* Angela Johnson, Moosewatch volunteer 2010, 2011, 2013, and 2017.

CHAPTER 12

HOW TO DO A BACKCOUNTRY OFF-TRAIL CSI

When a dead moose is found, we do a back-country CSI. The first thing we do is determine how the moose died. On Isle Royale, there are two primary ways that moose die: starvation or wolf kill. Moose do not die peacefully in their sleep. On rare occasions we might find a moose that died of some accident—for instance, my team once found a moose that somehow stumbled off a cliff edge and died from a forty-foot fall; we knew this because the moose's legs were rammed and buried straight down into the spongy ground about one to two feet. Moose will sometimes also fall into old mine shafts on Isle Royale and either die from the fall or drown, since many of the old mine shafts are flooded.

Death from starvation or wolf kill is easily determined. If you find almost every bone and they are all in the same general area within a twenty- to thirty-foot radius, it was likely a winter to early spring starvation—January, February, March, and April. If, however, you only find several bones scattered over a wider area (e.g., a back leg here and a front leg one hundred feet away) it indicates a wolf kill. With wolf kills, we rarely find most of the bones. With starvation deaths, we often find most, or even all, of the bones.

What can confound our searching for and finding bones is time. The longer ago the moose died, the more likely the bones have been buried by fallen leaves, trees, duff, and/or simply deteriorated away in the sun, the rain, and the snow. Given enough time, even a starvation moose will be difficult to find. All of the bones that are probably directly underfoot in the ground become invisible, courtesy of the forest's slow-motion burial.

Part of our CSI is doing an inventory of all the bones we can find and the level or degree the bones have been gnawed on. Even bones from a starvation will usually have some small measure of scattering and possibly gnawing from other scavengers (e.g., fox and ravens/crows). In some spots, like the middle of a swamp, finding a clearing large and dry enough to inventory the bones can be a problem.

We especially want to find the skull. There are a couple of reasons for this:

- Scientist can determine the age of the moose when it died by the wear patterns on the teeth.
- If there are pedicles on the skull—the base of where the antlers grow—we know it is a male, otherwise it is a female. Or, if the skull still has antlers, it is—obviously—a male.

We bring the skull and mandibles back to the scientist. If the skull retains its antlers, we bring that back too—grudgingly. Carrying an antlered skull off trail is heavy and difficult since the antlers grab and get tangled in everything.

When it is a male skull, we can estimate the month the moose died by the level of growth of the antlers or lack of antlers. Moose lose their antlers around December/January, so if it is an antler-less male, it likely died mid to late winter into the spring before antlers begin growing again.

The other bone we search for, and especially want to find, is the metatarsus. In humans, these are the larger bones in the foot, and in moose, evolution has repurposed the bone to be the lowest joint in their leg. The metatarsus is used as a proxy to determine a moose's health in vitro and for the first year of life.

Bone inventory. Counting the bones from a likely starvation moose. SOURCE: ALEC SMITH

We review all the other bones found too, especially examining the hip socket in the pelvis and the individual vertebrae; in these bones we're looking for osteoporosis/osteoarthritis. If found, we bring these bones in for the scientist. Any other bones with pathology (e.g., a broken bone) are also collected and brought to the scientist.

All of the bones are cataloged and photographed. After we've found, inventoried, and photographed all the bones, we also collect a lot of other associated information. We determine and log:

Bone inventory. Bones are lined up, cards are used to identify the group, and the find is sequentially counted. SOURCE: ANNA BURKE

- Exact location using GPS.
- The kinds of trees and plants growing in the immediate area.
- The nature of the ground (e.g., open, ridgeline, swamp, etc.).
- If it is a recent death, we will saw a whole leg bone in half and examine the marrow. If the marrow is watery, it indicates that a moose had used up its fat reserves and died of starvation. If the marrow is buttery or spreadable it indicates the moose was in relatively good health. For most wolf kills, however, all of the leg bones have been chewed to the point where the wolves got the marrow.

Perhaps one of the more important data elements we estimate is when the moose died. If it is a known wolf kill spotted during the Winter Study from an airplane, the exact date of death will be known. Moose with significant meat still on the bone obviously died in the recent past, and the year—and sometimes month—of death can be determined fairly easily.

We also examine the skull and sinus cavity for old maggot casings. Maggot casings, if found in an old moose death, suggests a 1996 starvation. The winter/spring of 1996 was a year when roughly 1,900—nearly 80 percent of all the moose on the island—died. At the time, the estimated count of moose went from 2,400 to 500. So many moose died that scavengers, let alone the wolves, couldn't eat all the moose meat that was available. And it fell to insects, flies, and maggots to take care of the job, leaving maggot casings in most 1996 starvation moose. Hence, we examine the depths of each older skull for maggot casings.

However, in most instances, once the meat is off of the bones, aging the bones can become dicey. Older bones can be sun bleached and cracked; they can be found in swamps or drier soil or under leaf duff. There are many environmental dynamics that can cause bones to age differently. For older bones—without maggot casings—I usually describe a five-year estimate of time of death. And I am fairly certain Rolf then uses the rest of our collected data and likely generates his own date of death, ignoring ours—or at least not putting too much stock in it.

All of the information we bring back to the scientists is then used to reconstruct the moose population through the years. For instance, if we

Temp Id #:
#'s of each bone found:

mtar ______ mcar ______
tib ______ rad ______
fem ______ hum ______
skull ______ mandible ______
scapula ______ sternum ______
sacrum ______ pelvis ______
ribs ______
Verts: cerv ______ thor ______ lumb ______
Total # bones (excluding ribs) with any flesh ______
(including little red flecks): ___________

Extent of wolf-chewed bones (especially the ribs):

How far were bones scattered:
COD, DOD, & other notes:

Vert Col intact? Y N
Skull attached? Y N
Mandible attached? Y N
Pelvis attached? Y N
Hide on skull? Y N
rear legs attached? 0 1 2
front legs attached? 0 1 2
How many hooves? _______

Specimens collected: UR UL LR LL Inc

skull pelvis sacrum hoof Mtar -or- Mcar

hide on legs: 0% 33% 66% 100%

Other hide Left: none some most

Three by five-inch data collection card (one side of a two-sided card). SOURCE: JEFFREY HOLDEN find a moose that we know died in the winter/spring of 2022 and from the teeth, it is known the moose was twelve years old at time of death, then the moose was born in 2010. The size, length, and weight of the metatarsus bone is then correlated with other environmental factors (e.g., temperature for the year of birth, snow depth, precipitation, tick levels, and so on). All of this data provides insight into what impacts the health of moose in their first year as well as throughout their lives.

I follow this basic process for any dead animals we find on the island, including foxes, beavers, or on rare occasions, wolves. If we find a wolf, we bring back every single bone we find no matter what. My groups have found several wolf skulls over twenty years and, in one instance, an entire dead wolf, or at least what the other wolf pack hadn't eaten. Dead wolves surprisingly don't weigh much and don't take up a lot of room in a backpack. They are much easier to carry than moose bones.

Anna's pack with a moose skull strapped onto it. SOURCE: ANNA BURKE

We frequently find dead moose that have been found previously; they're referred to as PDFs, or previously discovered finds. When this occurs, I usually complete an entire CSI write-up and review all of the bones we can find, and—occasionally—take a bone from the PDF on the off chance we've found a bone that was missed before. The reason to do a complete write-up is that the data we record changes and gets more detailed from year to year. If we go back far enough to around the year 2000 or before, the dead moose found then will not have GPS coordinates. Instead, the location would have been penciled in by a group leader on a map where they thought they were when they found the moose.

At the end of the backcountry CSI, we complete our notes and our photos of the "crime scene," and we load up bones, putting them into or strapping them onto our packs, and we keep hiking.

CHAPTER 13

HOW TO IRRITATE A MOMMA OWL

One wet, humid day, my group and I were slowly picking our way through the forest. We were at the northern edge of a little swamp with the requisite dead trees in it. There were a lot of bird noises coming from the swamp, so we stopped. I am awful at identifying bird songs, but somebody who is good at bird calls would have been in heaven or least challenged given the number of birds that were singing and chattering away.

I saw an owl in one of the tallest dead trees and said as much. The entire group stopped. Jeff Morrison, a dedicated photographer, was near me, and he said he saw the owl. Jeff was loaded down with a lot of camera equipment: his camera bag in front, a telephoto lens on his camera ready to take photos, and a full backpack. The camera gear probably added another twenty pounds to his load.

When I looked at Jeff, I noticed he and I were not looking at the same spot. But we both *saw* an owl. I definitely saw a large adult great horned owl in the large dead tree. I tried to follow where Jeff was looking, and he was looking at the ground to the left of the owl I was looking at. I looked where I thought he was looking and didn't see anything.

Then another owl flew into our nice little owl tableau. Our scene went all kerfuffle. An owl flew up from the ground—Jeff's ground owl. My dead-tree owl flew off in our direction and Jeff said, "Ohhhh," as he now saw the owl I had been looking at. Dead-tree owl flew directly over our heads. It was an owl rich-territory with intense owl eyes everywhere.

For a few seconds, the scene was shifting and confusing. I wanted to watch every single owl. Unlike a chameleon, I couldn't do that. I wasn't used to seeing owls in daylight, let alone three of them. They were gorgeous.

Our group was spinning around watching the owls, trying to take it all in. Jeff wanted to take photos (I suspect . . . I confess I was watching the owls) but I don't think he got any photos of owls in flight worth talking about. After a little bit, I decided we better move on. There was likely a nest someplace and we were bothering the owls.

Jeff wanted *dearly* to get photos of the owls. Being the decisive leader I am (really), I said no, and off we went. Let the owls have their space, and we'll keep hiking and find some dead moose, forgetting all about the owls.

We started hiking away from the swamp, dead trees, and owls. We headed west and were suddenly under a very large tree and a lot of guano. The ground was painted—unevenly—in a thick white paste. Bird shit, probably owl. *Oh, my boots.* Since we keep track of large raptors, including owls, I stopped the group so I could make notes about where we were, what we were seeing, and so on.

While I was taking notes, Jeff was wielding his camera, looking for owls, and snapping photos. I finished my notes and off we went, once again heading west with Jeff and his camera in the lead. Within fifteen or twenty feet, Jeff stopped dead in his tracks, camera up. *Snap, snap, snap.* Quiet please, photographer at work. I looked over his shoulder, and on the ground was the cutest, yet fiercest little owl: an owlet. Staring at us wondering what we were doing in their world. *Snap, snap.*

At this point, I noticed that we were not alone. Overhead was momma owl. Momma didn't like what she was seeing.

I told Morrison, "Hey, Jeff, it's time to go."

"Just a few more," he said.

"Uh . . . no. Time to go. Momma doesn't like us here." Overhead we were being buzzed by a large great horned owl with really big and

Owlet intensely interested in us. SOURCE: JEFFREY MORRISON

sharp-looking talons. My imagination was visualizing one, or more, of us getting some gashing wounds due to a talon-induced owl fury—a B-flick horror film filled with owl talons and blood.

The area of the forest we were in was a sort of open corridor in between trees, so the momma owl could make a long, deliberate, threatening path right over our heads, screeching all the way. I was suitably impressed and more than willing to leave. Jeff was taking photos. Meanwhile, the owlet was on the ground intently considering us, seemingly unconcerned. I finally got Jeff to leave, or, more likely, he had enough photos of the little cutie owlet on the ground. Off we went, hiking to the west.

In another twenty feet, we bumbled upon another smaller owlet on the ground. Jeff hefted his camera up—*snap, snap, snap*. Obviously two owlets (or more) had fledged from the nest, and . . . well, had not flown successfully, but floundered or plummeted to the ground. The second owlet considered us. Jeff clicked away.

Momma owl was unhappy. Another run down the tree corridor—somehow faster and more in earnest than previous runs—screeching. My imagination did a repeat of talons raking across my scalp—she was mad.

"Let's go, Jeff," I said with some urgency. I think he suddenly understood that we needed to go. The entire group set off with purpose. Everyone else in the group knew we were trespassing. This was not our space; it was owl space.

We hiked off . . . no bloody scalps (thankfully), but momma owl kept screeching behind us, slowly receding into the distance.

CHAPTER 14

BONEHEADED VOLUNTEER TRICKS

WHILE WORKING WITH THE WOLF-MOOSE PROJECT, I HAVE HIKED WITH about one hundred different people of varying backpacking experience levels. Many come well prepared and experienced. But not surprisingly, some of the volunteers have been naïve, inexperienced, ill-informed, and in some cases, simply boneheaded. This chapter covers a number of rookie mistakes one particular backpacker made once upon a time. . . . Here are some bits that I did or did not do my very first year I volunteered to look for dead moose.

I started hiking in 1976, back when there were only external frame packs, "ultra-light" didn't mean anything, and energy bars had yet to be invented. I hiked a lot from 1976 to 1987. There was then a fifteen-year gap in my hiking history. During this hiatus, backpacking gear and technology advanced a wee bit, but I kept and maintained my 1970s/1980s gear.

In other words, I came to the Wolf-Moose Project with what amounted to antique backpacking gear. Even so, I was a bonehead about number of things.

Old Gear—My Tent

I had gotten my tent in the early 1980s. It was a self-supporting dome tent, and I loved it. By the time 2002 rolled around, age, oxidation, and the poles (fiberglass) were splintering, making for excellent, nearly invisible slivers. The tent leaked a lot since the "waterproofed" seams had forgotten their mission and they didn't really put up much of a fight against the rain. It frequently rains on Isle Royale, often at night. A couple of nights I woke to little streams and ponds in my tent. The previous day's dirty socks were often used to sponge up the water. Lesson: Check your gear before you go.

Inappropriate Gear #1—Heavy Steel-Toed Boots

I remember thinking we would be off trail banging around trees and such, and I wanted to protect my feet, so I took some work boots. I still have these old work boots, and they weigh 4.6 pounds total, whereas my lightest pair of hiking boots weigh only 1.4 pounds total. This makes a world of difference, especially at the end of a day of off-trail bushwhacking. Crawling over rocks and downed trees is tiring and after a while, the leg muscles just don't want to do that anymore.

However, my work boots were broken in—they were and remain very comfortable, if heavy. A number of volunteers arrive on Isle Royale with brand new boots; boots that have *not* been broken in. Some end up with blistered, sore feet that are occasionally blistered enough where duct tape is the only logical, and perhaps best, backcountry medical treatment. Once, one of the people in my group with new boots wore them for an hour or two the first day, didn't like how they felt, and then did not wear his boots the rest of the week. Instead, he simply stashed his boots in his backpack all week and only wore Crocs while we hiked. Lesson: Test out your gear in situations as similar as you'll experience while backpacking to make sure they perform like you want them to. Boots especially need to be broken in.

Inappropriate Gear #2—Rain Gear

Backpacking rain gear remains a weird and difficult problem to solve. The best gear for water protection is the heaviest and bulkiest, which is not great for backpacking, and you'll likely sweat as much or more than the rain falling on you. The lightweight gear might not be robust enough, which is an issue hiking off trail. My first year, I decided I would just get the cheapest rain gear possible and live with it. Here are my notes on the rain gear that first year:

> I am . . . wearing my much-ballyhooed rain pants. I am not wearing the rain jacket over [my] sweater; as I know I would just sweat and I certainly don't need to create any more stink than I've already developed. The rain pants last for literally less than fifty feet of hiking. The waist-high hazelnut saplings ringing our camp do it in and the rain pants instantly have two huge gaping holes. After fifty feet, I am carrying a wadded-up bunch of torn and shredded rain pants, with a little fringe of plastic at my waist, all that remains of the pants.

The rest of that week was either rainy or hot and I was moist much of the time. Lesson: Rain gear remains a challenge—good luck with it, as you'll need to make compromises.

Plain Ignorance #1—No Flashlight or Headlamp

I'm not sure what I was thinking—or not thinking—about this item. I had no flashlight of any type with me. And just like everywhere on earth, Isle Royale gets dark at night. I do remember being thankful for the full moon and near full moon for much of the week. And Amy and Rich—other volunteers in my group—loaned me a head lamp. Lesson: Ask for help.

Plain Ignorance #2—Not Enough Water

I took my old water bottle that I'd had since I was perhaps ten years old: a cub-scout canteen that held about 1.5 liters of water. In many circumstances, it would be an adequate water supply. Unfortunately for me that first trip, we experienced unusually hot weather and we were primarily amongst swamps all week with little access to good water. On a couple of occasions, we would dig holes in a swamp and pump water from the resulting muddy puddle.

The photo, in which I am the one wearing glasses looking toward the camera, was from a stop one hot day when the rain had let up, but not the heat, and I remember as this photo was being taken that my legs were in danger of cramping due to lack of water. Also note, I am wearing blue jeans—a sure sign of ignorance when backpacking. Lesson: Our access—or lack of access—to water this first volunteer trip taught me to always have good water available for every campsite, and as a group leader, I do.

Wearing blue jeans and being nearly dehydrated during a break in 2002. SOURCE: JEFFREY HOLDEN

Plain Ignorance #3—Backpack Cover

This was something to cover your pack with when it rains. I had never used one before, didn't own one, and didn't know they were a thing. Phil, the group leader, pulled a large plastic bag out of the kill kit for me to use to cover my pack. These were plastic bags used for packing bones away as we crammed them onto or into our backpacks. Lesson: Repurposing gear, tools, etc. works. MacGyver lives!

For my first year on the island, I had gear that was sort of okay, but off-trail hiking was a different animal than hiking on trail. Each year I return to the island I usually have upgraded one or two items. Even now, after more than twenty years, I upgrade continuously.

That first year, I thought I was an experienced hiker, but I made quite a few mistakes. Reflecting on this has made me, as a group leader, a bit of a fussy old mother hen. A *forgiving* mother hen because forgetting gear or having the wrong gear is easy to do.

CHAPTER 15

EVIL JEFF

Every group leader gets their assignment for the week from Rolf Peterson. Usually, there are a handful of coordinates for known dead moose found during the Winter Study and perhaps some coordinates to check out previously found moose that were meaty when originally found and now are supposedly not so meaty. The coordinates tend to be grouped in an area so no coordinates are that far away from the general area a group will be in during the week. Often Rolf has more coordinates than can be easily hiked to for the week or they might be spaced a bit far apart. Better to plan too many spots to visit than not enough.

Rolf also attempts to group volunteers with similar capabilities. For instance, a group might have a number of people aged sixty and older or thirty-five and younger. The notion is that a group isn't composed of people who hike quickly with those that will need more time. Since many volunteers are new to the project, their abilities will be a guessing game. Volunteers do complete a questionnaire ahead of time to gain insight into the person's physical shape and backpacking experience.

All of this just means that at the beginning of the week, the route may or may not be well suited for the group. Weather can complicate things, and if we find a lot of moose, we won't have as much time to go everywhere

Rolf would like us to go to in his aggressive, ideal world. Occasionally there are injuries or people just wear down over the course of the week.

Part of the challenge for group leaders is laying out a route for the week to enable the shortest, easiest hiking without absolutely exhausting your group. This would mean days and distances were similar, there would be a good campsite in the evening with water and such, *and* we would go to every place Rolf would like us to visit. I find this logistical puzzle fun to solve.

Every group leader figures out their week in their own way. I tend to front-load extra mileage in the first two days of the hike, my rationale being that I'll do a tougher/longer segment of our hike early on to get it out of the way, I'll quickly see just what the group's potential is and know if I need to dial it back and, if so, how much, and if we're semi-aggressive, we might get everything done. I always factor in the weather—if I know we've got nice weather, I'll be more aggressive and more willing to hike full-pack off trail. If I know we've got rain coming, we definitely want to day hike; there is nothing so miserable as tearing down or setting up camp in the rain.

In 2010, two of my good friends, Angela and Dave, wanted to do the week on Isle Royale. They had never backpacked before, so I had helped them scope out gear and so on. I made sure they were in my group for the week. Others in the group were two other packing friends: Jeff Morrison (owlet photographer) and another David, as well as one of Rolf's undergraduate students named Scott. Scott was in exceptional shape; I immediately thought if we found an antlered bull moose skull that Scott would carry it, since antlered skulls are heavy and difficult to pack with off trail.

Our first day for this volunteer week, we got up early in the morning and took the boat to the island. Once on the island we traveled to Bangsund Cabin to gear up. Rolf gave me our GPS coordinates for the week. By this time, it was early afternoon. While the locations Rolf wanted us to explore for the week weren't overly large, they were in a remote and difficult-to-get-to part of the island, away from trails with interior lakes and more ridges to climb up and over than much of the island, which is saying something.

The day we hit the island was clear, bright, warm weather, but rain was coming. I planned our first half-day of hiking to cover a bit more ground than even I was used to. Everyone—other than Scott—would just need to suck it up.

At midday we started our hike from the Daisy Farm. Daisy is one of the major campgrounds on the south shore of the island. I planned our hike for the day to get to the north shore and Lane Cove that evening. If we were on trail, this would be a thirteen-mile hike. We, of course, were going to do a portion of the hike off trail to investigate a known dead moose. In the end, we hiked twelve miles that day, both on and off trail.

Immediately out of Daisy Farm, we were bushwhacking toward our first set of coordinates north and east of Daisy Farm into a river drainage called Tobin Creek. We were looking for a moose that a Park Service worker had found while hiking off trail.

At this point in the hike, it was a lovely day and everything was fine, but I knew we had a long day ahead of us.

After an hour or so, slogging through some swampy areas, we walked right up to the dead moose we were looking for and found a scattering of bones: a skull, some vertebrae, and a pelvis. We dropped our packs and spread out to find as many bones as we could. Very quickly, somebody in the group found more bones. I went to see what they had . . . another pelvis. Unless it was a very weirdly configured moose (i.e., one with two pelvises), then we had a second dead moose.

On occasion we find two moose at the same spot. It complicates things a bit because wolves spread the bones around and you need to sort them out and decide which bone belongs to which moose. If the moose were roughly the same size, it can be tricky. And if the bones have aged for roughly the same amount of time, there is another layer of confusion. But in terms of searching and getting dead moose counts, it is efficient; it's a two-fer!

We kept searching for more bones. And boy, did we ever find them. A third pelvis turned up. *Sigh.* At that point, I designated a central, relatively open spot as the location where we would sort the bones and inventory them. Bones came in, I sorted, hesitating a fair amount because—heck—I was not sure what went with what. I was staring intently at the ground and bones and somebody handed me . . . another pelvis. Pelvis number four. At this point we were laughing at the absurdity, but also wondering about the task of sorting them out.

Scott, the undergraduate student, started helping out with the sorting. This was nice because when he and I agreed on a bone and the particular moose pile it belonged to, it felt more correct. Frankly, Scott was better at sorting the bones than I was.

Eventually, a fifth pelvis was found. We had a five-moose pile-up. We ended up at this location for nearly three hours; a long time at one spot looking for dead moose parts. And it also ate into the time we would need to hike to the north shore. In the end, we had five moose, but only four skulls, about fourteen legs (both complete and incomplete), a lot of vertebrae, chewed ribs (scapula, anyone?), and those five pelvises. It took a long time to sort, inventory, and write my notes.

We finally put our packs back on and swore we would not find more bones, and even if we *did* find more bones, we'd deny seeing them. We still had about ten miles of hiking, and a good chunk of the afternoon was gone. The sun was out and it was hot.

We made it through the mildly swampy, squishy Tobin Creek drainage, up the southern face of the Greenstone Ridge, and onto the Greenstone Trail. The Greenstone runs along the spine of the most prominent ridgeline on the island. It is the main trail on the island going from the southwest end to the northeast end of the island. The trail on the section where we were at was along exposed rock with little shade. The rock had been in the sun all day, retaining heat, and radiating it back at silly hikers like us.

Somebody in our group had a thermometer and it registered over ninety degrees. We had about 1.5 hours of hiking to complete along the ridgeline in the heat, all with packs heavier than when we started. There were five sets of moose bones, so most, if not all, of us were carrying bones. In the heat, we were all dragging. We drank a lot of water. Those of us with extra water shared with those who were running out. My good friend Angela, not being shy, grumbled about the miles and the heat questioned what I was thinking, hiking this many miles on what was effectively a half-day—and literally the first time Angela and Dave had ever hiked before.

After the 1.5 hours in the heat, we were at the intersection of the Greenstone Ridge Trail and the Lane Cove Trail. Angela had renamed me—I was now "Evil Jeff" for planning such a hot and long hike. And we weren't done yet. We hung a left, turned north, and went down the Lane

Greenstone Ridge Trail in the sun looking north and east from Ojibway Tower. SOURCE: ANNA BURKE

Cove Trail for another two hours. The Lane Cove Trail was a steep, rocky trail and can be hard on the knees, especially when you're tired.

By the time we arrived at Lane Cove, it was approaching 8 p.m. and Angela wasn't the only one calling me Evil Jeff. We were weary as we set up camp, made dinner, and quickly fell into our sleeping bags. I was worried that Angela and Dave would both hate hiking and regret that they'd come along.

It began raining during the night and continued off and on for the next three or four days. During those few days we managed to cover good ground—those extra miles on day one helped position us for the rest of the week. The weather didn't always cooperate, though, and we did need to hike several days, full pack, off trail, in the rain. And then set up camp in the rain.

At one of those campsites, I put my tent up and wrangled a tarp between some trees so we would have a relatively dry place for dinner. Then I went to check on everyone to see how they were doing.

Scott Larson with an antlered skull on his pack. SOURCE: JEFFREY MORRISON

I checked on Angela and Dave last, wondering if they wished they were back home. I asked how they were doing, and Dave replied he was "having a great time" with a huge, genuine smile. Angela, too, was having a great time. She didn't, however, stop calling me Evil Jeff.

We put our miles in that week and found the dead moose Rolf wanted us to find, along with big, old antlered moose skull about as far from "home" as possible. Scott joyfully carried it. Midweek we had a break in the rain and we found an exposed rocky outcrop. We proceeded to cover this outcrop with all of our wet gear to dry. And we heard wolves howling multiple times, including one evening when we heard one pack to the west of our camp. They'd yip and howl. Then we would hear another pack, to our east, respond. That was a great evening.

Both Angela and Dave have since returned to the island multiple times; Dave ten or eleven times. They remain close friends. And I am still frequently referred to as Evil Jeff.

CHAPTER 16

THE SCIENCE OF THE WOLF-MOOSE PROJECT

How exactly do moose and wolf numbers impact each other? This is the central question of the Wolf-Moose Project.

In 1958 when Durward Allen and David Mech started the study, the main idea was to watch and count the moose and the wolves and see what they learned, and go from there. By early 2024, there were sixty-five years of moose estimates and wolf counts. At the start, the researchers believed there would be an equilibrium of the predators and prey; a yin and yang of opposite, but connected phenomena. They predicted that as moose counts increased, wolves would subsequently increase and reduce moose counts through predation, which in turn would result in reduced wolf counts because they wouldn't have as much to eat. A logical, but overly simplistic model.

A point of uniqueness to the Wolf-Moose Project is its longevity. Most studies depend upon limited funds, limited resources, and grants of perhaps three or five years in length. So a coordinated, well-funded project lasting for over sixty years is notable. Such longevity and continuity also underscores why shorter-term studies, while valuable, probably don't do their topics or scientific areas justice.

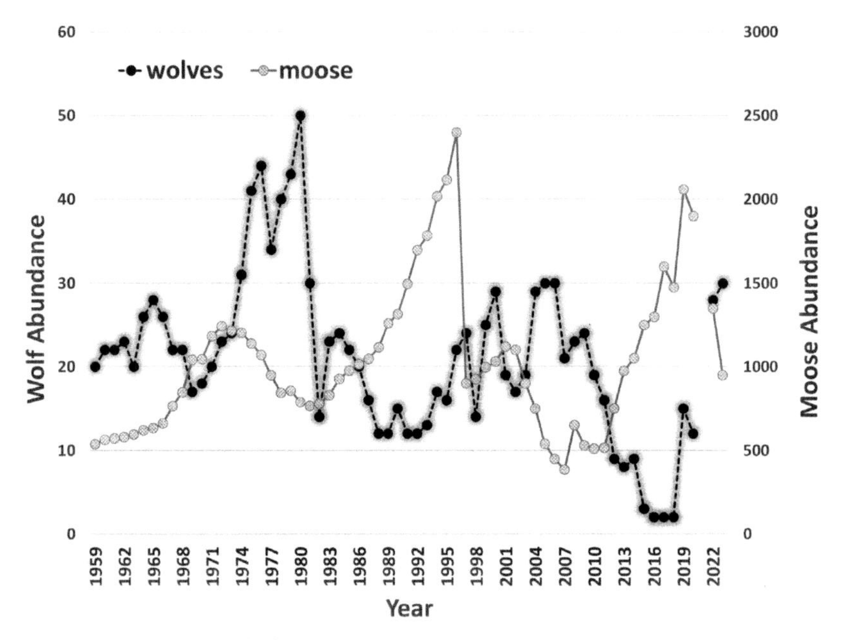

Wolf-Moose counts through the years. SOURCE: *Wolf-Moose Annual Report, 2022–23*, Sarah Hoy

The graph shows wolf and moose counts for every year since 1959. A quick look is easy to see that the two trend lines don't seem to have much to do with each other. A correlation of the data returns -0.3447, meaning there is a mild negative correlation (that yin-yang thing); when moose numbers go up, wolf numbers go down and vice versa. However, this is not a strong relationship between the two sets of data. If the data is offset by a year or two, the correlation gets stronger, but never strong.

While moose and wolves have an impact on each other, there is obviously something else going on. Or perhaps the data needs to be refined to incorporate nuance (e.g., counts of moose calves, prime-of-life moose, "elderly" moose). This search for more objective and richer data with additional dynamics to track is what has driven the research for decades.

This is why volunteer groups hike off trail every year, helping to expand the amount and type of data that would otherwise not be available. For instance, in 2022, volunteers hiked 678 miles and found 133 new dead moose for the project. Volunteers have been doing this type of work for over thirty years, putting in miles and gathering data and bones that the researchers, working alone, would obviously not be able to do.

The graph also demonstrates several time periods with extreme shifts in counts, usually decreases (e.g., the big die-off of moose in 1996, or wolf numbers dwindling to just two wolves in 2016–18). There are very few years in a row with "stable" or similar numbers. The graph presents no period of any length that looks like the period immediately preceding or following it. Both the moose and wolf numbers gyrate constantly.

The graph and numbers demonstrate that something else, or many something elses, is going on. One point is continually driven home with this research: nature is messy and unpredictable.

Interestingly, the study, as first envisioned by Durward Allen in the 1950s, was designed to last only ten years. But after those first ten years, Durward Allen—seeing the numbers that had been collected—found there were unanswered questions and continued the project. Another ten years came and went, and by 1980 the moose population had tripled in size, then declined by half, while the wolves more than doubled. The original "quaint" notion of a "balance of nature" was replaced by the idea that there were many rich, messy dynamics at play that impact moose and wolf numbers. And rather than thinking about shutting down the project and research, questions were asked that could only be answered by continued observation.

> The prevailing view of nature is that of a machine, whose rules are to be discovered by humans so that we may use nature as we wish. The most important scientific lesson from the wolves and moose of Isle Royale indicates otherwise. Isle Royale shows how nature is better understood as a series of inherently unpredictable events—each carrying consequences that unfold over many years. The reasonable response to that understanding of nature is for humans to relate humbly with nature. This lesson from Isle Royale would not be possible without have observed the dynamics there—so carefully and for so long.*

* John Vucetich, co-lead of the Wolf-Moose Project.

Those questions and further years of observation and data collection suggest a number of other important dynamics that impact, but still do not fully explain, wolf and moose numbers. These dynamics help make sense of what has occurred in the past, but might be of limited use in predicting future events. Some of those dynamics include weather, climate change, genetic inbreeding, tick outbreaks, disease and sickness, and random, unforeseen or unpredictable events.

Weather seems obvious in hindsight. By weather, we mean the day-to-day and month-to-month temperatures. Some basic "rules" as they apply to wolves and moose are:

- Hotter summers keep moose from feeding to some extent. The moose sit in the shade and chew their cuds, not eating, resulting in less fat for winter.
- Colder, longer winters means spring arrives later and moose—who are desperate for food come late winter into springtime—might not have the fat reserves to last, and they'll starve to death.
- Deeper snow means moose cannot find as much to eat in the winter. It also means it is more difficult for wolves to hunt, as moose move better in deep snow.
- Warmer summers and winters with high humidity tend to be good for ticks. Ticks need humid environments to thrive.

These four aspects demonstrate that even the weather's impact is varied and can set up or work in conjunction with other dynamics.

Related to weather is climate change. Long-term trends have shifted each of the four weather dynamics as well as reduced the likelihood of ice bridges between the island and the mainland. Ice bridges—or lack thereof—impacts the wolf population's periodic infusion of new genes.

Both the moose and the wolf populations on the island tend toward inbreeding because the island is remote and difficult to get to. With the recent introduction of new wolves to the island in 2018–19, the wolf population is not currently inbred, but prior to the new wolves, the population was severely inbred.

Ticks are increasingly making themselves part of the story. Ticks like the effects of climate change and appear to be doing very well on Isle Royale. Ticks can cause anemia and malnutrition in moose.

A disease that appeared on the island in 1980 caused the population of fifty wolves—an all-time high wolf count—to precipitously fall to fourteen by 1982. The cause for this drop in numbers was Canine Parvovirus, a highly contagious disease that also has a high mortality rate. The parvovirus was likely transmitted to the island from a pet dog, and infected the entire wolf population.

Random unforeseen or unpredictable events are the types of events that make sense in hindsight but cannot be seen or predicted ahead of time. For instance, a domestic dog bringing parvovirus to the island in 1980 is random and was instrumental in the wolf population dropping 72 percent. Another key random event was three wolves dying in a flooded mine shaft in the winter of 2011–12. This event was especially critical since one of the three wolves to die in the mineshaft was a female. Her death left only one breeding female wolf on the island and greatly increased the extinction risk for the island's entire wolf population. Random events do not occur frequently, but when they do, they can be significant.

The decades of observation, data gathering, and science underscores that we do not understand nature as well as we think we do. The Wolf-Moose Project started off with simply watching and counting wolves and moose and seeing what could be learned. Over time, more and more dynamics were identified and tracked. In the coming years, more phenomena are likely to be identified and tracked. We understand some pieces, but the idea that we are smart and know everything there is to know about nature is misguided hubris.

The more we know, the more we realize we don't know.

CHAPTER 17

TWO EXAMPLE STUDIES FROM THE PROJECT AND WHY LONG-TERM RESEARCH AND CITIZEN SCIENCE IS VALUABLE

THE WOLF-MOOSE PROJECT HAS NUMEROUS PEER-REVIEWED PUBLISHED STUDIES. The longevity of the project and the large number of researchers that have worked on, or with, the project over the years means the project is well known and has a high profile in the natural science world.

Year after year, dead moose are found, collected, and catalogued, making a valuable "library" of natural history. The resulting dead moose data are structured in such a way that a reasonable "census" of moose, on a year-by-year basis, can be reconstructed from the bones and data gathered in the backcountry by volunteers and scientists. This dataset is unlike any other in the world.

This means the project has completed a number of studies that might not otherwise have been possible *except* for the longevity of the project *plus* the contributions from the volunteers.

> Citizen-scientists have been a part of the Wolf-Moose Project for decades. The contributions of these volunteers have supported nearly every scientific revelation about the wolves and moose over the decades. They have

discovered hundreds and hundreds of wolf-killed moose and investigated their remains. Those investigations are the backbone of all we know about the moose population and essential context for understanding the wolves.*

What Moose Bones Teach Us about Human Disease

Humans are susceptible to many diseases. Often, studying diseases is difficult because of the lack of data over time—a longitudinal perspective—and the numerous confounding factors with human behavior (e.g., studies need to take into account factors like smoking, obesity, alcohol use, and so on). Analysis of shorter-term data, with numerous dynamics to account for, can lead to ambiguous findings.

Three human diseases of interest are periodontitis (gum diseases), osteoarthritis (cartilage degeneration), and osteoporosis (loss of bone density). These are all chronic age-related diseases that impact millions of people. Doctors and researchers have long thought there might be links between these three diseases.

A recent study by the Wolf-Moose Project, working alongside Duke University researchers, was able to assess possible connections of the three diseases with Isle Royale moose—wild moose—where many confounding risk factors (e.g., smoking) do not occur. The study determined the prevalence and severity of these diseases of over two thousand individual moose who died between 1959 and 2021.

The study found that female moose with periodontitis were 88 percent more likely to have severe osteoarthritis and more than twice as likely to have severe osteoporosis than female moose who did not have periodontitis. Results were similar for male moose: males with periodontitis were 60 percent more likely to have severe osteoarthritis and more than three times as likely to have severe osteoporosis than those without periodontitis. The study also found that male moose were more likely to get these diseases than female moose.

---

* John Vucetich, co-lead of the Wolf-Moose Project.

The study did not address or identify the underlying causes of these findings, but it is thought the bacteria that causes periodontitis encourages inflammation and tissue destruction, which may exacerbate osteoarthritis in joints. These bacteria also can impair cells in the body that grow and heal existing bones while simultaneously increasing the activity of cells that break down bones and lead to osteoporosis.

The research adds to a growing body of evidence suggesting periodontitis increases the risk of developing other serious health conditions including diabetes, strokes, cardiovascular disease, and pneumonia. It also highlights the importance of developing and maintaining good oral hygiene habits.

A related earlier study from 2010 found a connection between poor nutrition in early life and arthritis in adult moose, adding to research that links the likelihood of arthritis in humans to low-quality diets.*

Lead and Mercury in the Atmosphere

Air pollution in decades past used to be a bigger deal in part because the pollution was visible to the naked eye. The early and mid-1970s saw passage of the Clean Air Act and the removal of lead from gasoline. By the early 1980s concentrations of mercury and lead had dropped in Eastern North America significantly.

It was, however, still difficult to assess mercury and lead contamination in specific ecosystems because both mercury and lead are impacted—not by region-wide levels of pollution, but rather by local point sources (e.g., individual factories).

---

* Sarah R. Hoy, John A. Vucetich, Leah M. Vucetich, Mary Hindelang, Janet L. Huebner, Virginia B. Kraus, and Rolf O. Peterson, "Links between Three Chronic and Age-Related Diseases, Osteoarthritis, Periodontitis, and Osteoporosis, in a Wild Mammal (Moose) Population," *Osteoarthritis and Cartilage* 32, no. 3 (March 2024): 281–86; Cyndi Perkins, "What Studying Moose Bones for 65 Years Can Teach Us about Human Diseases," *Unscripted Research Blog*, Michigan Tech University, January 16, 2024.

Isle Royale is an ideal place to measure such phenomena because there is no local point source nearby and because Lake Superior acts as a buffer. Any mercury and lead on the island would have originated from 120 miles away. Any declines noted on Isle Royale would mean there was a region-wide decline in mercury and lead and not just a single point source.

Unfortunately, nobody monitors either mercury or lead on Isle Royale. However, the levels of these two heavy metals were recorded in the teeth of moose. Mercury and lead were deposited from the atmosphere onto vegetation, which in turn was consumed by the moose, and a trace amount made its way to the teeth of moose and permanently locked into the enamel. The study used teeth of moose who were known to have born between 1952 and 2002.

The study's findings were that mercury and lead levels were high and steady until the early 1980s when the mercury concentration dropped suddenly by about 65 percent; it has remained steady at this level since then. Lead levels began to drop in the early 1980s and continued to drop for the following two decades until lead concentrations were 80 percent lower than they had been prior to 1980.*

Both studies provide insight into areas that are not directly connected with the project's original goal, devised by Durward Allen and David Mech, to simply count wolves and moose. These studies underscore the value of the long-term perspective of the Wolf-Moose Project with decades of data being discovered and collected often by volunteers. Science is a process about trying to understand the world we live in and to continually ask questions.

---

* John A. Vucetich, P. M. Outridge, Rolf O. Peterson, Rune Eide, and Rolf Isrenn, "Mercury, Lead, and Lead Isotope Ratios in the Teeth of Moose (*Alces alces*) from Isle Royale, U.S. Upper Midwest, from 1952 to 2002," *Journal for Environmental Monitoring* 11, no. 7 (2009): 1352–59; "Moose Teeth Record Long Term Trends in Air Pollution," Wolves and Moose of Isle Royale, https://isleroyalewolf.org/node/41.

CHAPTER 18

VOLUNTEERS SUPPORT OTHER SCIENCE EFFORTS

THE MAIN THING THE VOLUNTEERS DO WHILE ON THE ISLAND IS LOOK FOR dead moose. But as long as there are volunteers on the island collecting data off trail, they might just as well collect data for other projects or gather more information for the Wolf-Moose Project, right? Tasks groups have done, either once or sometimes frequently, include gathering data for antlers, snowshoe hare, gray jays, tick activity, active beaver dams, other animals like fox, martens, and weasels, raptor nests, wolf locations, balsam fir heights, granite, and beaver foraging distance (BFD).

Just like deer, every fall/early winter, moose lose their antlers—what we call "sheds." When we find them, we GPS their location and measure the size of the pedicle (a fifty-cent word meaning the base of where the antler attaches to the skull). Then we mark the antler so that any subsequent group that finds the it will know it has been found before. We mark the antler by using a bone saw and putting an X on its base.

Data collection for snowshoe hares is quick. Anytime we see a snowshoe, I simply note the date that we saw one. At the end of the week, we tally the number of snowshoe hares we saw per mile hiked. At the end of the research year, each groups' data is aggregated and an overall snowshoe hare number per mile hiked is generated. This indicates the relative abundance of the hares from year to year, which varies greatly.

Pre-1900 snowshoe hare would have been one of the primary prey of Canadian lynx on the island. The lynx no longer exist on the island—this is because of humans hunting, not wolf predation.

Snowshoe hare are a minor secondary prey for the wolves; generally, 90 percent or more of a wolf's diet on Isle Royale will be moose. The remaining portion will be hare and beaver.

Similar to snowshoe hare, we simply note when we see a gray jay. Their numbers also vary greatly from year to year. I am unsure why we track gray jays. I do know I love them, as they are inquisitive, intelligent, and feisty.

One year, as our group prepared to leave the island, I was rearranging my pack off to the side. Something hit me on the head and I thought it was one of my clever backpacking friends tossing a pebble or something. I eyeballed my friends, who were perhaps thirty feet away, trying to determine who the culprit was. I remember thinking they were all doing an excellent job of keeping a totally innocent straight face. I went back to my pack and was hit on the head again. My friends still remained totally innocent looking. And then I noticed a gray jay in the tree next to me, about fifteen or twenty feet away, looking me over, cocking its head from side to side, and I swear it was smiling. The gray jay was waiting until I wasn't looking to fly over and bop me on the head for fun. I love that playfulness.

We gather data on tick activity by doing our best to take detailed photos of the flanks of moose. From the photos, researchers can estimate the amount of fur lost to ticks during the winter and by extension the level of tick prevalence during the previous winter. Ticks, as we know, have an outsized impact on moose.

The island has many lakes, streams, and therefore quite a lot of beaver activity. As we hike off trail, beaver dams are a main "highway" we use to get from one side of a swamp/lake/pond/stream to the other. When we know a beaver dam is active, we GPS its location.

In recent years, when the wolf population count was greatly reduced, the number of active dams increased significantly, resulting in a number of new ponds, swamps, and little lakes. On occasion, trails need to be re-routed and/or have boardwalks installed (i.e., elevated walkways).

If we find the bones or remains of any of the other "large" mammals like fox, martens, and weasels on the island, we treat them similarly to a dead moose. We write up the find and bring back bones, especially the skull and teeth.

When we find an active raptors nest (e.g., eagles, osprey, owls) we note the location with GPS. This information is passed along to the Park Service largely to steer other hikers away from the nests. Active nests are easy to identify since the ground underneath the nest is carpeted in guano.

Large or weirdly shaped/unusual antlers—We gathered large or weird antlers one year, and that was more than enough. That year, when we found a large or unusual antler, we picked it up and packed it out. We find *large* antlers on Isle Royale fairly frequently, and the one year we were collecting them, we found many more than I wanted to find. At the end of the week,

The obligatory holding antlers up to our heads for the 2022 end-of-week group photo in Rock Harbor. SOURCE: ANNA BURKE

my backpack was approaching seventy pounds, or at least twenty pounds more than I like to carry.

When the researchers do not have a good handle on the location of wolves, either individual wolves or packs, we sometimes go out in search of them. We don't want to actually *meet* them and have a confrontation; the scientists just want to know where they are, the shape of their territory, and perhaps where their den is. How do we find the wolves? Two primary ways: investigating known wolf dens and checking the telemetry for collared wolves.

The first is to investigate known wolf dens, which are known from prior years. We are given coordinates of known wolf dens from years past and we hike to see if there is any activity around them. I've visited a number of former wolf dens and never found any recent activity. I am always amazed at just how small the dens are. Wolves, while top predators, are not really all that big and can curl up into the smallest, coziest little spaces.

Telemetry for collared wolves is the second way to locate wolves. If wolves are radio collared, we sometimes take a telemetry unit into the field and attempt to find them. The telemetry signal is "squishy" in that its strength and direction vary depending on how close or far the wolf is. The closer the wolf, the stronger and more directional the signal is. If the wolf is far off, the signal will be faint and cover perhaps a ninety-degree swath of land in a particular direction. When this occurs, you must hustle to another point several miles away and attempt to get the signal again and then triangulate. Of course, if the wolf is on the move, it will have covered more miles than you and complicate any efforts at triangulation.

When wolf numbers were reduced in the 2010s, the moose population swelled, and in the winter, starving moose ate most every balsam fir they could find. For several years we would record data, GPS coordinates and count the number of trees for *any* balsam fir that exceeded five feet in height. This did not happen very often; most balsam fir were nibbled on and were less than five feet tall, an indication that moose were stressing the environment for food and that starvation deaths in the late winter/early spring would be more likely.

Information about granite is in support of another non-wolf-moose research effort that involves glaciation from the last ice age. The major ridgelines on Isle Royale have elevations at or above nine hundred feet so any non-volcanic rock on ridgelines might be granite carried there by glaciers. As with most data, we GPS and photograph it, and pass the information along.

Another data-gathering activity when the wolf numbers were reduced was Beaver Foraging Distance, or BFDs. Rolf seemed quite pleased with this acronym. BFDs are the estimated distance a beaver gets away from their dam and lodge, the notion being that during normal wolf predation, beavers will not venture far from their dams and the safety of the water. When the wolf numbers were down, the distance the beaver ventured away from their pond likely increased.

We would document the distance by finding recently chewed trees as far away from an active dam as we could and then determine the distance. This was usually determined by simply counting off paces; a bit cumbersome in the off-trail, fallen-tree, swampy environments we would be in, but do-able. We found distances of 200 to 250 feet—this was quite a distance away from the safety of water, meaning the beaver on the island for a year or two in the 2010s were not overly concerned with wolves.

Since the reintroduction of wolves several years ago, the BFD distances have dropped *and* the number of active beaver dams has also gotten smaller. All due to wolf predation.

Recent beaver activity and the active dam it was near. The group is using the dam as a way to get across the swamp. SOURCE: MEGAN HORODKO AND ANNA BURKE

What do the scientists do with this data? If it isn't directly related to the Wolf-Moose Project (e.g., granite rocks at or above nine hundred feet in elevation), the scientist will pass our data along to other scientists. Some data we gather is for simply building—or contributing to—a baseline that might take years to develop (e.g., the balsam fir height). Other data is more directed (e.g., Beaver Foraging Distance) and will develop into an academic paper fairly quickly. And some data is simply gathered and analyzed to see if there is anything there (e.g., I don't think the large antlers led to much, aside from, perhaps, confirming that moose on Isle Royale are smaller than mainland moose. I do know it confirms that I do not like carrying that much weight).

As a volunteer, I consider myself an informed amateur and I have a reasonably good grasp of the basics and perhaps some insight into the project. Obviously, I am not steeped in the project and all of its work. Even so, I love being part of the project and contributing year after year.

CHAPTER 19

THE LODESTONE INCIDENT

OFTEN THE HIKING DAYS ARE LONG. DAYS CAN STRETCH OUT BECAUSE THE spots you need to check are a long way away. You might find more moose or other items of interest than you thought you would, and you have to do multiple backcountry CSIs, which can take time. Or the hiking might be just plain hard. Weather gets in there too.

In the middle or late afternoon, people might start dragging. If we have been day hiking, then everyone starts thinking about being in camp because we know where we will be for the night—up ahead of us someplace. *Are we there yet?* Or if we haven't decided on where to camp, we just want to find a spot and say that is the end of hiking for the day.

One day my group and I had such a long, tiring day. It was springtime of 2011 and we were day hiking, knowing we'd be back at Moskey Basin Campground in the evening. We had spent the entire day hiking off trail, finding and investigating several sets of coordinates southwest of Angleworm Lake.

It was getting late in the afternoon and had been off-and-on raining all day. But we'd done what we had intended to do for the day and we were heading back to Moskey Basin campground where we had our gear in a shelter. Many of the campgrounds had three-sided shelters and because of the rainy weather, a shelter sounded very good.

Our route back to the campground was fairly easy. Campgrounds are found along trails, and trails are long ribbons that cover a lot of ground, which meant we didn't need to have pinpoint navigation accuracy to get us back to camp. All we needed to do was stumble onto the trail and know which way to turn, and we would eventually get to camp.

That was how I was navigating at that point in the afternoon; kind of laissez-faire. We were tired and it was raining. There wasn't as much talking as usual. The sun was behind the rain clouds so we needed compasses to navigate. But because precision wasn't required, I was not checking my GPS or compass as often as I do.

We were supposed to be following a bearing that was south-southeast. We schlepped along, tired, damp, hungry, and looking forward to the shelter at the camp. I checked my compass, and the needle was all goofy, not pointing in the direction I thought it ought to be. It was moving around like a tiny invisible hand was turning the needle.

I stopped the group for a quick break to grab some water or a handful of peanuts. I checked the compass again. Still all goofy. I didn't like that.

Angela (*Evil Jeff* Angela) came up to me and could see I was confused. "What's up?" she asked. I was hesitant to admit the needle of the compass was goofy, but I showed her my compass. The needle was still wandering around in an arc of perhaps sixty degrees.

"Oh," she said. Angela got out her compass and checked it. Also goofy, but a different goofy because it was pointing in a totally different direction. Dave came up with his compass. We now had three compasses to refer to, and all three were exhibiting the same meandering-needle phenomena and finding "north" in different directions.

We laid our compasses on my map to ensure they all were on as flat a surface as we could get. The compasses persisted in doing the same thing. I said, "I don't like that." Part of my brain was thinking, *What the hell?*

Angela suggested we move from where we were standing and go up a small little hill that we were next to. We climbed to the top of the hill and checked our compasses again. No goofy, all normal. They all were pointing solidly in the same direction. We weren't lost after all.

We knew how to get out and home to the campground, and we had also found a magnetic lodestone. What fun. Not what I needed, but interesting.

We continued hiking, found the trail, turned right, and were shortly at Moskey Basin Campground and in our shelter. The shelters at Moskey were right on the water, so the views were excellent. The rain then really started coming down. We rested, did a few chores haphazardly, prepared and ate our dinner, and watched the rain.

That evening's entertainment was watching an osprey fish in Moskey Basin directly in front of our shelter. The osprey would hover, dive, and come up empty. This happened repeatedly, perhaps eight or ten times. We were hoping the osprey would come up with a fish at some point, but it never did. Even so, a great show.

CHAPTER 20

WHAT I HAVE IN MY BACKPACK FOR THE WEEK

WHEN YOU ARE CARRYING FORTY-FIVE POUNDS, GIVE OR TAKE, WHAT DOES THAT mean you have in your backpack? Backpack sizes are, for some goofball reason, listed in liters. For a week-long hike, a fairly large pack is necessary. Sixty liters or larger tends to be what works best; I have two packs at, or above, seventy-five liters. Anything smaller and you'll end up with gear on the outside of the backpack. If you're only hiking on trails, gear on the outside of the pack is not a big deal. But looking for dead moose means you will be off trail scrambling through the underbrush, over trees, and so on, so gear on the outside of the pack can inhibit progress or—worse—simply disappear at some point. There's nothing like getting to a spot where you will camp for the evening and finding out you don't have your tent poles or tent stakes. Both—unfortunately—have happened to people I've hiked with.

Everyone has a tent, a sleeping bag, and a sleeping pad (e.g., a thin insulating "mattress" both for comfort as well as to help keep you warm). And, of course, the backpack itself. The pack can be heavy or less heavy, but rarely light. If the backpacks have a large capacity for longer hikes *and* are robust (e.g., for off-trail hiking), they tend to be heavy, weighing over

Some of the major items we carry in our backpacks. SOURCE: JEFFREY HOLDEN

five pounds when empty. Over the years I've continually upgraded these four main items, trying to cut weight while at the same time increasing my comfort. Of late, I've gotten the weight of these four items to just seven pounds, five ounces total. This is considered "ultra-light" gear, and it makes me giddy knowing how light it is. Ten years ago, the same four items would have totaled more like twelve to fourteen pounds. Shaving off five to seven pounds might not sound impressive, but as the miles pile up while hiking, the lighter load is very noticeable. Your feet and knees will thank you.

You want to have clothing, but not too much. You do *not* want to have multiple changes of clothing, but rather layers. When I hike, I have one pair of pants for the week (pants that get very dirty), a couple of T-shirts, a sweater (brown, of course), a couple pairs of underwear, and more socks than anything.

> The best advice . . . was to bring a warm pair of sleeping socks to be worn only in your tent. After three days of steady rain, my socks saved my morale

> and kept me going. They were the only item of clothing I had with me that was dry and didn't smell like moose.*

A thin, puffy jacket (e.g., goose down or something smashable and warm) is necessary. All clothing should be ripstop. Nothing should be cotton; quick-drying synthetics are best. Off-trail backpacking is not a fashion show and clothes just take up space, weigh a fair amount, and they'll be dirty once you put them on and get about thirty minutes of hiking under your belt—sometimes less time than that.

Camp shoes are very important. It is lovely to get out of your hiking boots at the end of the day and let your feet have a breather in something lighter and more comfortable. Fortunately, camp shoes do not need to be fashionable. I wear Crocs—they are god-awful *ugly* but effective camp shoes.

Rain gear is always a compromise. I admit, I've never quite figured out rain gear. Good rain gear is heavy and bulky, so most backcountry hikers do not have "good" rain gear. Backpacking rain gear is usually light and good for *repelling* rain for perhaps thirty minutes or maybe even an hour, less if the rain is heavy. The better the rain gear is at repelling the rain, likely the "better" it is at not breathing and then making you sweat. You'll sweat to such an extent that you get nearly as wet from sweat as you would the rain. Fortunately, long-lasting heavy rainstorms are rare, and non-cotton clothing dries quickly. As with all clothes, the rain gear should be ripstop, otherwise it will shred to pieces.

Everyone carries food—for themselves as well as communal food. In the groups I lead, I make dinners ahead of time so that we're only cooking one dinner in the evening instead of multiple people making multiple different dinners. My dinners are semi-famous—at least in looking-for-dead-moose circles—since they are lightweight, easy to make, and tasty. Most people will carry ten to fifteen pounds of food at the beginning of the week. Hikers tend to eat about 1.5 pounds of food daily for breakfast, snacks, lunches, and dinners. If you have more than two pounds of food per day, you will likely return at the end of the week with several pounds of food uneaten (i.e., you were carrying too much).

---

* Pattie Evans, Moosewatch volunteer 2024.

Other important communal gear includes

- A tarp—Tarps are great to put up in case it rains. Nothing is quite so miserable as attempting to make dinner in the rain without protection. Determine where you will camp for the night and put up the tarp even if you don't think it will rain, because every once in a while, it rains when you don't think it will. And, should it rain when it doesn't look like it will, and you've got the tarp up, you look like a genius. That means I've looked like a genius exactly twice in my life.
- Water filters—Water is important stuff, so you need at least two water filters of some sort. We all carry a lot of water—probably four pounds or half a gallon most of the time—and we need it constantly for drinking and for cooking. However, we don't need it for cleaning (sorry). That means pumping or filtering water. Water is sourced from Lake Superior (the best water), inland lakes (not bad water), or creeks (also not bad), and if you have set up camp someplace not so nice, you might be drinking swamp water. Filters take care of swamp water, but it dirties the filters quickly and makes for slow filtering.
- Stoves and fuel—You need hot water to make coffee in the morning. Real coffee, not instant. And to make hot water, you need a stove. Therefore, there are usually two stoves in each group. Coffee is important. I make coffee by using individual coffee filter packs or "pillows" and throw them into a pot of water, and Bob's your uncle. The stoves are handy for dinners too.
- First aid kit—Knock on wood, my groups haven't used the first aid kit much. We mostly use some bandages for scrapes, moleskin for blisters, and ibuprofen for the obligatory aches of hiking, but because you're on your own—and a long way from anywhere—the kit is fairly comprehensive. Most groups will have a first aid kit from the project and many people will bring their own medicinal items. Ibuprofen is never in short supply. The project's first aid kit has plenty of bandages in case of—I don't know what sort of ghastly injury would require the use of so many bandages. There are also instructions for broken bones, and ointments and unguents. There are no IVs or defibrillators, but even so, a fairly large amount of medical stuff takes up space and weight in somebody's pack.

- Kill kit—This is the generic name we use for all the tools and equipment for collecting moose bones and anything else we might be gathering information on. Again, we don't always use all of it, but we have it just in case, and it takes up space somewhere in the group.
- Navigational equipment—When I first started volunteering at the turn of the century, that meant a compass and a map, since GPS technology was just coming into wider use. No GPS meant that every single step off trail was with full packs. GPS has made navigating vastly easier and more precise. Hiking can now be done with day packs instead of full packs. Of late, smart phones with downloaded maps are getting into the mix and providing another mode of figuring out where we are and where we might want to go.

Most people have items for hygiene (e.g., toothbrush), headlamps and spare batteries, and a pack cover for rain.

If after getting everything into your backpack you still have space and the inclination to carry more weight, you can. Extra stuff—stuff that isn't absolutely required—that people bring in their backpacks include books, notebooks, cameras (assuming they aren't using their phones), or a towel, as some of us jump into Lake Superior, which is mighty chilly most of the year.

Odd things that people have brought along over the years include a change of clothes for every day of the hike; chairs (there are such things backpacking chairs, but these were bigger and more for tailgating); a large lantern, which isn't necessary since most of us are dead asleep before it gets totally dark; and an espresso maker that was never used during the week so I have no idea if it worked well or not. Mostly, though, people forget one or two items they wished they had brought. That's where sharing amongst the group comes in.

CHAPTER 21

HIKING IN SNOW

The first weeks of every season on Isle Royale have few visitors on the island, in part because the several boats that service the island have yet to begin their "regular" schedules. The boats come to the island about half as often as they do later in the season.

This is also the time of the season when the Park Service opens the campgrounds and hikes all the trails, noting where crews need to pay attention to falling trees and washouts. Everything at Rock Harbor, including the lodge and restaurants, generally do not open until Memorial Day. Isle Royale, while officially open, is yet to be fully engaged for the season—it is just waking up from its winter slumber.

Volunteer groups start coming onto Isle Royale during this first part of the season in early to mid-May. Weather for these earliest groups can be cold, rainy, snowy, and sometimes just plain unpleasant. However, there is no ground cover to speak of and that makes looking for dead moose much easier.

Some years have a lot of snow, and 2013 was one of those years. The snow and cold weather lasted through winter, into spring, and then kept on going. By the time early May rolled around, no boats were going to the island due to snow and ice. Team I, the first group of volunteers, was canceled. Team II, scheduled for the middle of May, got to the island, but had to contend with ice and snow everywhere. The volunteers had a difficult time moving around.

I was leading a group for Team III that year in late May. Before I got to the island, the figurative alarm bells were ringing. Rolf, on the island with the Team II volunteers, wanted all of us in Team III to have snowshoes because Team II did not have snowshoes and apparently could have used them. I volunteered to coordinate the snowshoe effort. Rolf pointed me to a resource at Michigan Technological University, an on-campus group that provided sporting equipment, including snowshoes, to students.

I reserved sixteen pairs of snowshoes. Thankfully, I had a truck, so that amount of gear, while significant, could at least be stowed in the truck bed with no problem. As I traveled north to Isle Royale, I stopped in Houghton at the Michigan Tech campus to get the snowshoes. The rental place was a large ramshackle house on the edge of campus. Naturally, nobody was at the rental place when I arrived. I went and got lunch, ate it, and then came back and—*whew*—somebody was at the rental place. I Loaded up the truck and continued on to Copper Harbor.

Getting onto the Isle Royale Queen, the boat that travels to Isle Royale, can be a confusing, jumbled time: it is early morning, the caffeine has yet to kick in, and there are lots of people milling around trying to ensure their gear is all packed neatly and on board and that they make it on the boat too. Throw in sixteen pairs of snowshoes, and it gets more complicated. I worried the snowshoes would get bent or broken and be useless and then be a problem returning. But it was all fine in the end, and nothing broke or got lost.

We boated across Lake Superior. Before we got too close to the island, we could see snow remnants. Normally, we would dock and disembark at Rock Harbor, but much of Moskey Basin and Rock Harbor was still thick with ice. We ended up disembarking at a non-standard smaller dock, but it was ice-free. Docking and disembarking at the smaller dock was different, interesting, and not particularly efficient, but heck, we were on the island.

We then ferried from the dock to Bangsund Cabin in Rolf's little fishing boat. This took several trips. Eventually all of the volunteers, gear, and snowshoes were at Bangsund. We did our normal gearing up, making sure we had what we needed food- and communal gear–wise, got our marching orders, and we were set to go.

By this time, I'd also seen that, while there was significant ice mostly everywhere, it didn't look to be deep enough to actually require snowshoes. I've winter backpacked both with and without snowshoes, and I generally think the snow needs to be at or above your knees before it is worth the hassle of snowshoes.

Rolf was sending us west to Moskey Basin and into the interior of the island toward Lake Richie and beyond. We had this route because a volunteer group from the prior team had attempted to go this way but was prevented by the snow and ice, and never even made it to Lake Richie. Now it was our turn to give it a try.

Normally, Rolf would have taken us by boat to the Moskey Basin Campground and we would have hiked from there, but the Basin itself was iced-in, so we hiked from Daisy Farm to Moskey Basin Campground first. Then we could begin our week.

As we hiked, there was plenty of ice and snow, but again, nothing that required snowshoes. Sometimes the trails would have drifts, and for ten

A scruffy red fox spotted along the trail from Daisy Farm to Moskey Basin Campground at the end of winter. SOURCE: ANGELA JOHNSON

or twenty yards we would have to contend with fairly deep snow, but we'd plow on through with little problem besides our boots getting wet and cold. For a portion of our hike, we were escorted by a slightly mangy looking fox.

At some point early in the hike, Ron Eckoff decided to put on his snowshoes and see if they helped. He strapped on his snowshoes and off we went. Ron reported they were difficult to use if there was no snow, and, as if to highlight this observation, he tripped and fell several times. Getting back up on your feet with a backpack and snowshoes is difficult. And, for those of us who weren't in that predicament, kind of funny. After a short while, Ron took off his snowshoes and that experiment was done. Nobody else tried snowshoes that week.

We made it to Lake Richie and stashed all of our snowshoes, as we decided they weren't going to be helpful and nobody wanted to carry them around. Lake Richie at the beginning of the week was almost completely ice covered with just a few open spots here and there along the shore.

For the rest of the week, we hiked and camped as we normally would do. Frequently, we hiked in snow, but it was rarely difficult. Most lakes were iced over. One brilliant sun-shiny afternoon, we postholed up to our hips in snow on the north side of a slope for around twenty minutes—this was the only spot during our entire week where snowshoes would have been useful.

We wondered how many dead moose we might have missed buried under the snow and ice. Weirdly, we saw more snowshoe hare than I had ever seen before. Their fur was a mix of their winter and summer coats, a sort of dirty white—exactly what a lot of the island look liked with melted, muddy snow.

Toward the end of our week, we came back to Lake Richie to camp. The lake was almost completely ice-free by this point; we had a family of otters in the water immediately in front of our camp for entertainment. We got our stashed snowshoes, strapped them on our packs, and hiked out, ending our week of snow hiking.

CHAPTER 22

HYGIENE

WHEN BACKPACKING—OR ANY KIND OF CAMPING, REALLY—HYGIENE TAKES A hit, and compromises must be made. During a seven- or eight-day off-trail hike, even more compromises are made, and they all add up to one stinky week. Actually, it is never as bad as I fear it might be, but it isn't a clean week, that's for certain.

First, there is one hygiene activity that everyone continues with no interruption: brushing your teeth. Get a toothbrush, a little tube of paste, some water (although water isn't required), and you can keep the teeth clean, white, and shiny. Lord knows you wouldn't want to have bad breath in the backcountry.

After brushing your teeth, most hygiene activities are either ignored or severely cramped. In other words, "clean" doesn't really happen. There are no showers or baths. In Isle Royale's backcountry, we have access to lakes and to Lake Superior. Lake Superior has super-clean water, but it is also super cold most of the year. In the springtime to early summer, Lake Superior might claw its way up to forty or forty-five degrees. Turn on the cold water at home in your sink and that is roughly the temperature of

Jumping off the dock into McCargo Cove in 2023. SOURCE: ANNA BURKE

Lake Superior in the spring. Getting into Lake Superior is not a leisurely extended activity.

Inland lakes, while warmer, are not warm. They also frequently have leeches, a silty, squishy bottom, and lots of branches and sticks. Just getting into a clean spot in the lake can be difficult.

The photo shows four of us jumping into Lake Superior at McCargo Cove; the resulting "swim" lasted perhaps ten or fifteen seconds for each of us as we jumped in, surfaced, in some fashion verbalized—loudly and with vigor—our observations that the water was frigid, and then swam like heck to get out.

A backpacking "shower" can be done by pulling out a stove, heating water to around one hundred degrees, filling a Nalgene with the heated water, and going off into the woods for a sponge bath. Not great, but very doable, and worth the time.

Soap—nope. Leave No Trace discourages soap; even biodegradable soap needs to break down, and in theory it should only be used two hundred feet from water sources, which sort of defeats the purpose, making it inconvenient.

"Clean" is mostly not going to happen unless you can handle cold water. Otherwise, you need to accept that you are going to marinate in yourself all week. Some people try to combat this by applying a lot of deodorant, but that seems to be a losing game. You either smell like you times ten, or you reek of deodorant. By weeks' end, you and the rest of your group are a malodourous bunch. Hand sanitizer, however, is very popular, especially where food preparation is concerned.

Often as we hike, we'll sometimes be near a trail. When that occurs, we take advantage of the easier hiking on the trails. Trails lead to campgrounds, and when in a campground, there are outhouses. Backcountry camping makes outhouses attractive; truth be told, Isle Royale outhouses are, relative to most outhouses, pretty clean. If outhouse locations are known, we might go slightly out of our way to visit them.

Otherwise, in the backcountry, you poop and pee in the woods. You will have to find a private spot away from the tents, dig a hole four to eight inches deep—Isle Royale as a rocky place makes finding spots with soil that deep difficult—and do your business. You will have to carry your used toilet paper out (Ziploc bags are useful, *double* Ziploc bags). Pro tip—when locating your cat hole, make sure you find ground cover nearby with large, robust leaves. Those leaves do the dirty work (ahem), and then less toilet paper is required to handle the finishing touches. And, when you hit an outhouse, you can get rid of your used toilet paper there.

Shaving is especially easy to give up as far as everyone is concerned, except for one dedicated volunteer. I only mention shaving because once, somebody in my group shaved religiously every morning. Dick Murray managed to dry shave each and every morning with a safety razor. Brave man. Nobody else shaves. Ever.

Changing into clean clothes during the week is another semi-futile gesture; most people do not pack many clothes for the week. Should you put on clean clothes, they will simply be dirty and sweaty in short order. You are dirty and smelly and your clothes will match you by the end of the week.

Once our week of finding dead moose is complete, we return to Bangsund Cabin and clean up. Bangsund Cabin does not have running water, so the end-of-week bath consists of a large bucket of Lake Superior water heated inside Bangsund Cabin on a propane stove, a ladle, and a bar of soap. A line of other Moosewatchers wait their turn to get clean. Candy

Peterson corrals and cajoles everyone to get their baths because we all have to be clean and pretty for dinner. Once it is your turn for a bath, you grab your warm water, ladle, and soap and carry it back into the woods a short distance where there are large blue tarps draped in the trees to create two shower stalls. You stand on duckboard and ladle warm water on yourself, soap up, rinse. And repeat two or three times. It is lovely.

CHAPTER 23

MY GROUP IS ON-SCREEN TALENT FOR A TELEVISION SHOW

THE WOLF-MOOSE PROJECT IS FAIRLY WELL KNOWN, AND IS COVERED BY magazines, newspapers, radio, and television shows. In 2014, a Canadian film company did a series on wildlife in and around the Great Lakes. They planned to do one episode on Isle Royale and the Wolf-Moose Project, and a portion of that episode would cover the volunteers and their work hiking off trail looking for dead moose. My group that year ended up being the one the film company worked with. We had about five minutes of screen time, which was fun.

The cameraman for the company had a lot of heavy equipment, over one hundred pounds. Because of the weight, the film crew did not follow us during the week, or even for one day. Instead we came to them at the end of the week.

Several months prior to this, a mature male moose wandered into Daisy Farm, immediately across Moskey Basin from Bangsund Cabin. The moose then died right in the middle of the campground, at which point Rolf and Candy came and, with some help of campers, dragged the moose out of the campground. Moose are big, therefore the dragging only got the dead moose about thirty or forty feet beyond the last outhouse in the campground. A rotting moose would be unpopular in the campground and might attract wolves, so it was best to keep them out of the campground.

Back to the film company: The plan was that my group would come back a day early from our off-trail wandering and then use this dead moose, at least what was left of it, and *pretend* to look for it for the cameras. We would be the onscreen talent.

We returned at the end of the week in the rain. My group went to Daisy Farm's pavilion, not too far from that last outhouse, and stowed our gear to keep it dry. The camera guy told us what he wanted us to do as he filmed us pretending to look for moose bones.

We went to the bones and they were not scattered. The scavengers that got the meat were likely raven/crows, fox, and bugs. Not wolves—wolves scatter bones. But the cameraman wanted us to "find" bones from a wolf kill. Therefore, the first thing we did was disarticulate the bones and spread them around. Then, once we realized where the heavy camera was going to be set up for much of the shoot, we moved some of the bones because the outhouse would be in the view—not the wilderness experience they wanted to film. We put moose bones hither and yon, under trees, in grass, in a wet damp squishy area—all over.

It rained hard, so we retreated to the pavilion to wait it out and discuss what exactly the film company wanted us to do and say. They wanted us to just act naturally and say what made sense. At this point, one of my group decided he did not want to be involved and left to go look at some mine shafts several miles to the east. Rather than having five of us look for bones, it would now be four.

The rain slacked off and into the deep, deep backcountry we went, forty feet from the outhouse. The cameraman and director started suggesting things for us to do and things to say, all of which were things we would *not* do or say. After several iterations of this, they finally relented and let us pretend and do our thing, ad-libbing the whole thing.

I'd like to think my ad-libs were fabulous and informative. Actually, at first, I think we were all pretty stilted as we yelled at each other. When you are in the backcountry looking for bones, you yell at each other, since you are rarely next to each other. While we were filming, we would pretend to be looking into the distance and then yell to the other searchers who would, in reality, be about ten feet away. Then the camera would pan to whomever was "in the distance," and there we were.

We laughed a fair amount, which, naturally, was edited out of the final version. The filming showed each of the four of us finding a variety of bones and bringing them back to our "staging" area. Everyone got a chance to "find" bones and talk a little on camera, mostly yelling, "Bones!"

Eventually, we laid the bones out for easier inventorying (e.g., leg bones with leg bones, all the ribs in a pile, etc.). I knelt down so I would be in the camera view along with the bones, and yammered on a bit about how it was obviously a wolf kill because the bones had been dragged all over a spread of two hundred feet, *and* the bones were chewed on quite a lot. I did note, in my yammering assessment, that the moose was pretty healthy—not too old, the teeth were fine and not worn down much, and there wasn't any sign of arthritis in the hips and therefore was unusual that the wolves were able to take a bull moose down.

Probably in reality, it had just starved.

We ended our camera time with me throwing a bone on a bone pile, several of our group saying, "Let's go," and me then giving a bogus bearing of 210. And we all walked back into the woods.

Given the nature of the staging and us totally winging it, the final footage—originally broadcast in Canada—was pretty decent. The following year, one of the volunteers in my group said he had seen the show and had never heard of the Wolf-Moose Project, but he had to come out and volunteer.

CHAPTER 24

MORE ABOUT MOOSE

Most years, the moose counts on the island are in the 500 to 1,500 range with some years well above and some below those thresholds. This range makes Isle Royale one of the highest-density spots on the planet for moose, with 0.22 to 0.65 moose per square kilometer (or 0.6 to 1.7 moose per square mile). As volunteers hike off trail, there is a pretty good chance you are going to see a moose on any given day. Even if you aren't off trail, but in the campgrounds, you have a good chance of seeing moose.

The highest moose count on Isle Royale occurred in 1996, when there were 2,500 moose (1.08 moose per square kilometer). At the time, the scientists on the island knew this count of moose couldn't be sustained.

In the late 1800s, there were no moose on Isle Royale. Sometime around 1900, some moose—count unknown, but probably not many—jumped into Lake Superior and thought, "I am going to swim to that island off there in the distance." And moose have been on the island ever since. Given that moose are nearsighted, I wonder if they could see the island, or perhaps they smelled it. Regardless, they jumped in and completed a fifteen-mile swim to the island.

Isle Royale is toward the southern geographic limit for moose in North America. Moose prefer colder weather and dislike heat. During the middle of the day in summertime, they rest in the shade, chewing their cuds. Generally, moose are found north of forty-five-degrees latitude—think mostly in Canada with a few incursions into the United States in New England, the upper Midwest, and along higher elevations in the Rocky Mountains.

If there weren't predators, ticks, and deer, the geographic range of moose might extend further south. Deer prefer warmer climates than moose, which is why they are to the south of moose. The two species do not share much geography. A main reason for this lack of overlap is brain worms, which sounds like something from a bad 1950s science-fiction movie.

Brain worms are a parasitic roundworm that affect deer, moose, elk, caribou, and other deer-like creatures. White-tailed deer are the normal host for the brain worm. For some reason, brain worms do not affect white-tailed deer greatly, but they have a serious impact on moose and elk. The worms eat into the brain and spinal column and cause all manner of bad symptoms (e.g., weakness, lack of coordination, circling, deafness, impaired vision, and paralysis), ultimately leading to death. Thus, deer and moose do not usually occupy the same geography. Bullwinkle and Bambi will *never* be best friends.

Moose Eat and Chew, Chew and Eat

Isle Royale moose are big—weighing at six hundred to one thousand pounds for adults—and are solitary members of the deer family. Interestingly, even at this weight, the moose on Isle Royale are amongst the smallest moose on the planet. This is an example of the island effect, also called Foster's rule. The concept is that larger creatures become smaller when food resources are limited because of land area constraints (e.g., being on an island).

All moose spend a lot of their time either eating or chewing their cud. Like cows, moose have a four-chamber stomach with the largest chamber called the rumen. To get the most out of their food, moose eat it, digest it some, regurgitate it, chew on the cud, and re-swallow it for a second round of digestion. Moose patiently chew their cud about eight hours every day.

The moose "social" calendar is limited. Moose—like deer—rut in September/October, bull moose lose their antlers around December/January, and calves are born after about an eight-month gestation or around May. The rest of the time they are eating, sleeping, chewing their cuds, and/or avoiding wolves. During the warmer months, they eat as much as they can to increase their body weight by as much as 25 percent to put on fat so they can make it through the winter months.

Through the winter, their diet is severely restricted and the only plants available to them on Isle Royale are balsam fir and, when really desperate, the bark off of trees (e.g., aspen). The balsam fir are an indicator of both the number of moose on the island and the level to which they are starving in the winter. Many balsam firs are only five feet tall or shorter, but might be twenty or twenty-five years old due to moose winter browsing; basically bonsai balsam fir. Because of the lack of easily available food in the winter, moose pass much of the time resting and ruminating, hungry and waiting for spring. For about four or five months of every year, moose lose weight every single day.

Because they eat trees and sticks, a problem that moose encounter is getting wood stuck into their mouths and jaws. A severe occurrence can lead to necrosis and infections of the jawbone, which in turn can lead to death. The photo shows one old bull moose with a stick wedged in his upper palate; the moose died of malnutrition since the stick interfered with eating and rumination.

Once winter is over and spring slowly approaches, some of the first real food available to moose is water shield, which are like little lily pad–type plants. There are a number of inland lakes on Isle Royale where water shield is common, and in April and May these lakes are busy with moose.

On several occasions, I've seen over twenty moose simultaneously at Lake Ojibway, all totally ignoring humans and focusing on eating the water shield. They'll walk within ten or twenty feet of you to get to the lake, and you better not be in the way.

The aquatic plants are fairly rich in protein and sodium. There aren't a lot of options for sodium on the island. Aside from the aquatic plants, moose make use of a number of salt licks (i.e., they essentially eat salty mud) around the island. Eating aquatic plants also means the moose are safe in the water because wolves do not go in the water. Interestingly, moose have

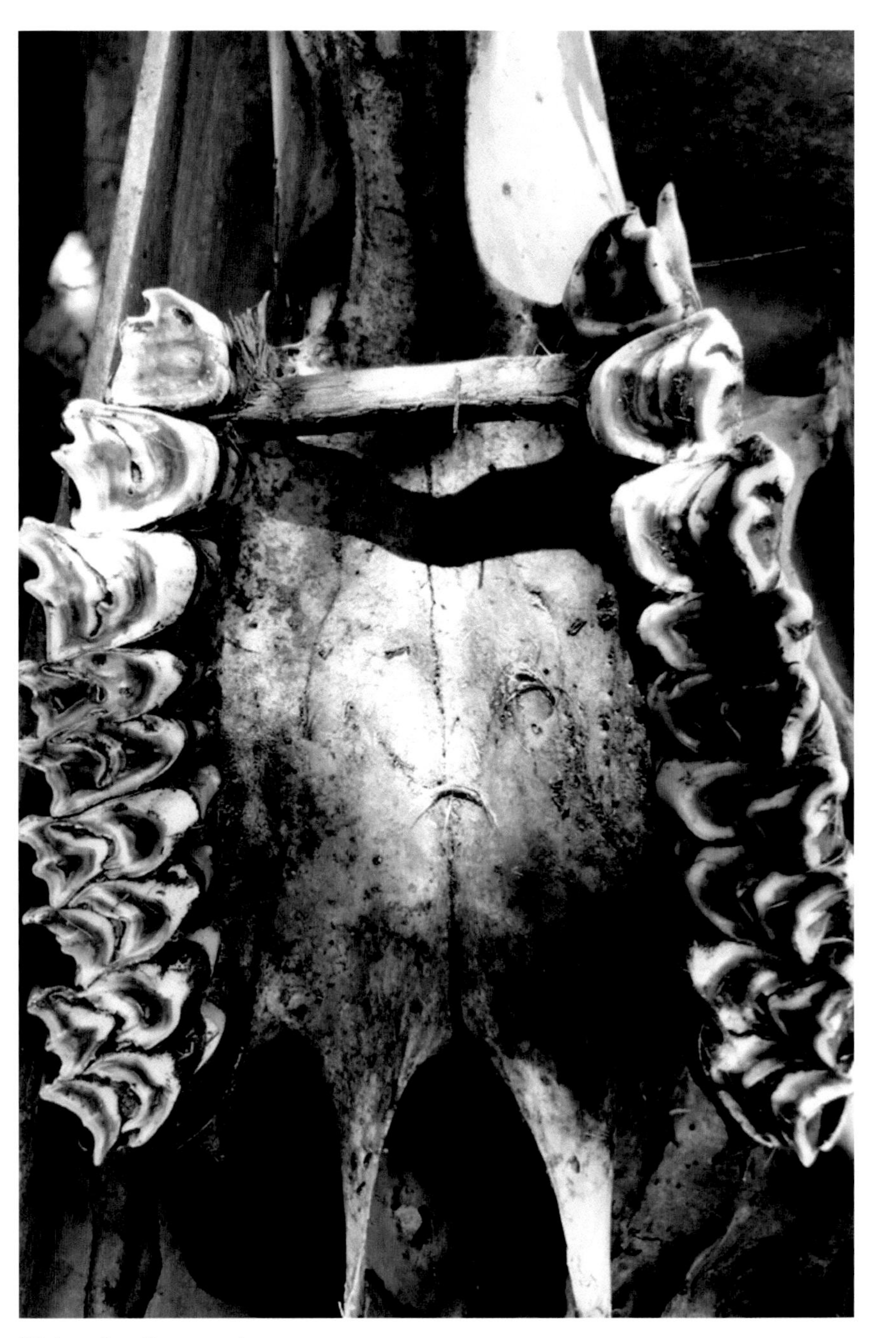

Stick wedged in moose jaw. SOURCE: WOLF-MOOSE ANNUAL REPORT, 2020–21, ROLF PETERSON

Collared cow feeding in a pond in the springtime. SOURCE: MEGAN HORODKO

nostrils that can close, or pinch off, so while they forage underwater plants, they don't get water in their noses—one of their many superpowers. The photo shows a bald collared cow feeding on water shield in the springtime with, undoubtedly, nostrils pinched close. She is bald due to tick activity during the winter months.

Once spring and summer is in full swing and real plants are growing, the moose diet expands considerably. They'll still eat last year's water shield that has sunk to the bottom of the pond, as well as other aquatic plants, flowering plants, twigs, leaves, pine needles, cones, and fruit.

Because their diet swings wildly, in the winter, moose poop looks like large, dry rabbit pellets. These pellets are found *everywhere* on the island. Come summertime, their poops all looks like cow flop. A tale of two diets.

Chapter Twenty-Four

Moose and Ticks

Moose were new to North America in the last ten to fifteen thousand years, with recent arrivals coming over from Asia and Siberia via a land bridge. This had significant implications for moose, since when they arrived in North America, the moose really hadn't encountered ticks before. Deer, as a comparison, have been dealing with ticks for a zillion years. Because of this, deer are reasonably good at grooming and keeping their tick problems to a manageable minimum.

Moose, being new to the tick game, are wretched at grooming for ticks. Moose essentially ignore ticks until they have a serious problem. This often occurs when it is cold and the moose's fat reserves are being depleted. This is when moose learn what anemia and blood loss mean. Imagine one hundred thousand or more ticks all over you, each having a nice little blood meal at the same time. For those of you keeping score at home, that amount of blood, at 0.5 milliliters of blood per tick, totals the equivalent of over one hundred sixteen-ounce bottles of Coke.

An engorged tick and ticks on moose. SOURCE: ROLF PETERSON

That is a lot of blood. And the 0.5 milliliters are, by the way, on the low side of a moose tick's capacity. Fully engorged, a moose tick will be about the size of a small grape; a small grape with eight little wriggly legs.

When they do it, moose groom ticks by basically scraping the ticks off by rubbing against trees. Moose will scrape and scrape until they are bald at that spot. Done often and frequently enough, by springtime, if the moose is still alive, they might be approaching bald everywhere.

The image shows an adult moose with a patch of skin, black in color on the shoulder hump where the ticks would have been a problem. The

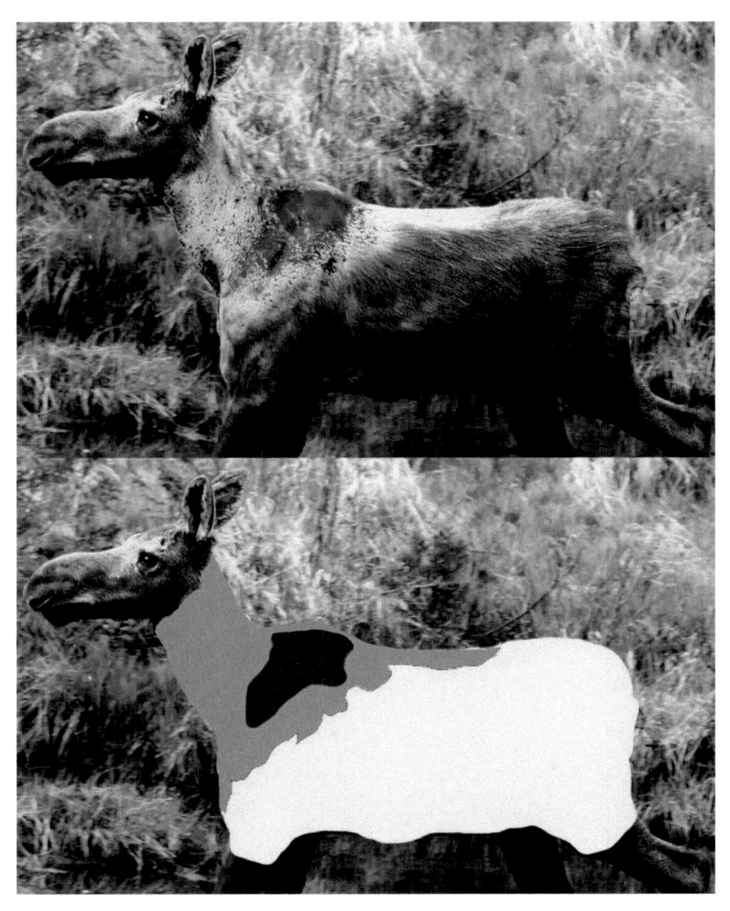

Moose in springtime with tick damage. SOURCE: LEAH VUCETICH

neck, chest, and some of the back shows lighter bits of broken tick-damaged fur, with the rest of the fur healthy with little or no tick damage. Relatively speaking, this is not a large amount of tick damage. Some moose, come springtime, will show significant tick damage over much of their body.

Ticks can, and do, kill moose, either directly through blood loss or indirectly through exposure, anemia, and disease. Ticks are good vectors for disease, like Lyme disease, that humans get. Thankfully, the ticks on Isle Royale do not carry Lyme disease.

Research suggests warmer summer temperatures may accelerate tick egg development and increase egg survival. For moose, warmer summers mean they are less active trying to keep cool and the moose might not build the fat reserves they need for winter. Warm summers are a double whammy—the ticks like it; the moose, not so much.

Ticks do not like cold, harsh winters; they prefer warmer winters. The Wolf-Moose Project has been tracking winter tick prevalence since the 1990s. With climate change, the ticks are doing well, meaning their numbers are increasing—and that the moose are having tougher winters. This is occurring not just on Isle Royale, but all across North America.

CHAPTER 25

WOLVES AND GENETIC RESCUE 1

As of early 2024, I've hiked nearly one thousand miles on Isle Royale and have only once seen a live wolf for perhaps two seconds from a distance of about one hundred feet. I did not get a good look. However, even when you don't see them, you know they are there. You can hear them howl, and they seem to love leaving scat in the middle of trails and especially on boardwalks over swamps. There are plenty of paw prints, and on several occasions you might get up in the morning and see one of your own footprints with a wolf print smack dab in the middle of it, so you know they were sniffing around your campsite in the night. I *love* that.

As volunteers, we do not actively look for dead wolves, because there are not that many wolves on the island, and when they die, they are much smaller than moose and their bones do not last as long. So randomly finding a dead wolf is unlikely. On the rare occasion a dead wolf is found, we pay attention.

Wolves are relatively new to the island. They arrived on the island in the late 1940s or 1950s by crossing over an ice bridge from Canada. Through the mid-twentieth century, ice bridges formed fairly frequently and so wolves, or any animal, could conceivably walk to and from the island on the ice in the winter.

In the 1960s, ice bridges formed, on average, two out of every three winters. However, since 2000, ice bridges only form about once or twice every ten years. The declining frequency of ice-bridge formation is likely the consequence of climate change, reflecting warmer winters and especially windier conditions. Wind inhibits ice formation. Even when ice bridges form now, they do not last as long as they did in decades past.

The decline in ice is significant because it reduces the possibility of a wolf immigrating from the mainland, which DNA studies suggest are necessary for maintaining the genetic health of the Isle Royale wolf population. An ice-free Lake Superior means the family of wolves on the island would increasingly only have siblings or cousins to mate with.

Because of wolf immigration and ice bridges becoming less and less likely, scientists have been noting and warning of inbreeding in the wolf population on the island. Inbreeding makes itself known in a number of ways, but especially in the vertebrae of the wolves. The vertebrae become lopsided and malformed through the generations. By 2000, roughly one-third of the island's wolf population suffered from a malformity known as lumbosacral transitional vertebrae (LSTV). By contrast, only one in one hundred wolves suffer from LSTV in healthy populations. You don't need to be a statistician to know those two rates are wildly divergent. The photo shows an example of an Isle Royale wolf with four misshapen sacral vertebrae, especially on S1. This wolf likely suffered damage to the nerves that control its tail and hind legs.

Then in the late 1990s and early 2000s, in spite of the concern for inbreeding, the wolves became more robust and territorial for a string of years. The 1990s was also when the Wolf-Moose Project began systematically gathering wolf scat with an eye toward DNA analysis. The project couldn't afford the cost of DNA analysis for over a decade, but when analysis of the scat began in 2009, there was a surprise in store for the scientists.

The Old Gray Guy

The photo shows the Middle Pack on the ice; each pack on the island is named and the Middle Pack was claiming the middle section of the island

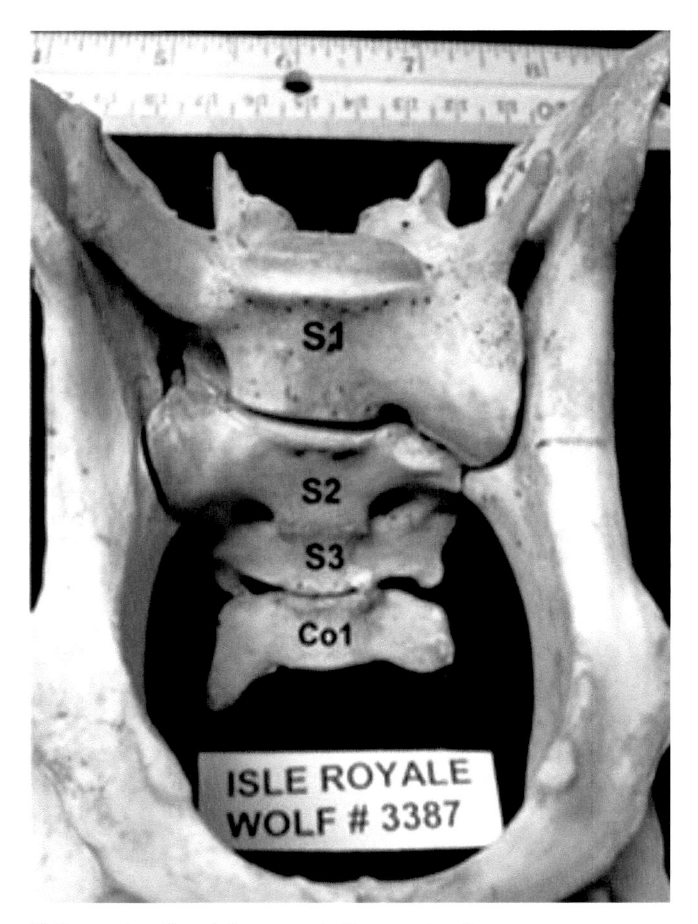

Malformed wolf vertebra. SOURCE: WOLF-MOOSE WEBSITE, JOSEPH BUMP

for a time. The largest, lightest-colored, and handsomest wolf in the photo is the Old Gray Guy. So what is so special about the Old Gray Guy?

The scientists noted the Middle Pack was taken over by a new alpha male sometime between February 1997 and February 1998. This alpha male was designated as Wolf #93. Wolf #93 was no ordinary wolf. He was distinctly larger than other Isle Royale wolves. His strong territorial behavior completely pushed out the West Pack, driving that pack to extinction in 1999. With Wolf #93 leading, the Middle Pack grew to ten wolves by 1999, the largest pack size observed on Isle Royale in almost twenty years.

The Old Gray Guy. SOURCE: WOLF-MOOSE ANNUAL REPORT, 2010–11, JOHN VUCETICH

The scientists also noticed Wolf #93 turn very light in color, almost white, as he aged. At that time, they dubbed him "the Old Gray Guy." While turning light with age is not uncommon among wolves in general, this had never been observed before with Isle Royale wolves. In subsequent years, the scientist also noted three other whitish-colored alpha wolves. All of these observations were reported before any wolf poop DNA analysis had occurred.

In 1999, during a Winter Study research flight, the scientists observed Wolf #93, the alpha male of the Middle Pack, defecate on a frozen lake. When the pack left the lake, a scientist landed the plane and collected the scat. The scientist could directly tie that scat sample to Wolf #93.

Several years later, when funding for DNA analysis was available, Wolf #93's scat was analyzed. The scientists noticed Wolf #93 was different. Wolf #93 first appeared in the poop record in 1997. Wolf #93 carried DNA markers that indicated the wolf was an immigrant from Ontario—likely on one of the rare ice bridges that formed. And the other light-colored wolves on the island were his direct descendants.

Wolf #93's story on Isle Royale represents an important opportunity to better understand genetic rescue. Genetic rescue involves introductions of one or more unrelated individuals into an inbred population as a means of mitigating the problems caused by inbreeding. This is a potentially important conservation concept and tool. But genetic rescue is not well understood because the opportunities to closely monitor an isolated population before and after a known immigration event are limited. For that reason, Wolf #93's, or the Old Gray Guy's, immigration to Isle Royale represents a special opportunity.

A "successful" genetic rescue would show an increase in a population's vital rates after immigration. However, the evidence for increased vital rates in the Isle Royale population was not clear cut. There was no statistically detectable difference in several key metrics the scientists use to monitor wolf predation and reproduction success. Coincident to the time the Old Gray Guy was active, the moose numbers on Isle Royale declined dramatically in response to food shortages, severe winters, and tick outbreaks. A clear response to the Old Gray Guy's immigration event might have been disguised by these confounding factors.

What was known was that the Old Gray Guy's genetic constitution was superior to that of native Isle Royale wolves. He soon chose to mate with a wolf that shared half of his genes; he sired twenty-one offspring with his own daughter. Two of these offspring began breeding with each other and they established a new pack in 2007 (the Paduka Pack). In 2003, the breeders of the East Pack were siblings who had been born to the immigrant and an unrelated Isle Royale wolf. In other words, within several years of arriving on the island, five of Isle Royale's six breeding pairs were either the Old Gray Guy or one of his offspring. All told the Old Gray Guy was an alpha wolf for eight years from 1998 to 2006, giving birth to thirty-four direct offspring, who in turn had many offspring.

The Old Gray Guy was perhaps too successful for the benefit of the population. His rapid and dramatic success, and that of his offspring, led to quickly rising rates of inbreeding by 2003. On Isle Royale, the last wolf unrelated to the Old Gray Guy died in 2007. By 2009, 56 percent of all the genes in the Isle Royale wolf population traced back to the Old Gray Guy.

The Old Gray Guy likely died sometime during 2006 to 2007 at age ten or eleven, which was old for a wolf in the wild. As with most wolves, he simply disappeared and was never found after he died.

At one time, scientists thought Isle Royale wolves had avoided inbreeding depression despite being isolated and highly inbred. The early 2000s discovery of bone malformities (i.e., lumbosacral transitional vertebrae or misshapen vertebrae) suggests the wolves hadn't avoided inbreeding depression. The discovery of the Old Gray Guy's identity through DNA analysis indicated the wolves haven't been quite so isolated as once thought.

Science initially had some of the major features of the wolf population all wrong. But science is a process and, as with all good science, the new data informed the situation and contributed new information about the wolves of Isle Royale. The Old Gray Guy's story is a reminder of the rewards of long-term research and continually asking questions and exploring.*

---

* Jennifer R. Adams, Leah M. Vucetich, Philip W. Hedrick, Rolf O. Peterson, and John A. Vucetich, "Genomic Sweep and Potential Genetic Rescue during Limiting Environmental Conditions in an Isolated Wolf Population," *Proceedings of the Royal Society B* 278 (March 30, 2011): 3336–3344, https://royalsocietypublishing.org/doi/10.1098/rspb.2011.0261.

CHAPTER 26

THE 1996 BIG MOOSE DIE-OFF

IN 1995, THERE WERE TOO MANY MOOSE ON ISLE ROYALE. THE ISLAND IS approximately 210 square miles, of which roughly 80 percent are attractive to moose, meaning there is something for moose to eat. A portion of land in the middle-west of the island experienced a month-long fire in 1936 where roughly 20 percent of the island burned. The fire was so hot, it burned the island down to the bedrock and what has grown back since is largely not of interest to moose. This middle section of the island has a lot of aspen and birch which is lovely for hiking on and off trail, but not so interesting for moose, and hence, wolves don't spend a lot of time there. This is gradually changing as more moose-friendly things grow there, but to a large extent, this is not moose territory.

The two maps show where the moose are more commonly found (dark areas, lower map) and where wolf tracks are more commonly found (dark areas, upper map). And, not surprisingly, the maps look similar; wolves go where the moose are.

So while the island is quite large, not all of it is attractive to moose. The moose-functional territory on the island covers roughly 165 square

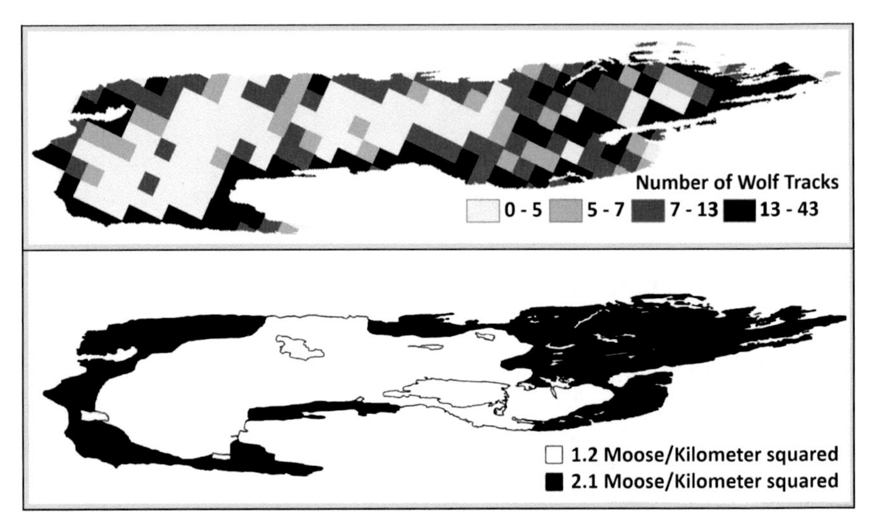

Moose and wolf track locations. SOURCE: WOLF-MOOSE ANNUAL REPORT, 2011–12

miles. In 1995, there were an estimated 2,400 moose on the island, or about fourteen moose for every square mile of moose-functional territory. These are rough calculations, but no matter how you slice the math, this was a lot of moose; probably more moose per square mile than any place on earth at the time. If you had visited the island in 1995, you would have likely seen multiple moose.

Why so many moose? It can, in part, be explained by low wolf numbers at the beginning of the decade; wolf population counts were in the low to mid-teens for much of this time period. Probable severe wolf inbreeding as well as the low counts resulted in fairly low Kill Rates. Through the early 1990s, the moose population grew, nearly tripling to the estimated 2,400 in 1995.

But parties do not last forever and the good moose times came crashing down. Scientists knew a day of reckoning would come to the moose at some point because they had seen this play out before. In the 1930s, prior to wolves migrating to the island, a similar significant moose die-off occurred and was monitored by Adolph Murie, a famous naturalist and wolf researcher. The cold winter of 1933 was the first phase of a moose

Moose nibbling balsam fir. SOURCE: WOLF-MOOSE ANNUAL REPORT, 2013–14, GEORGE DESORT

die-off and the fire of 1936 added a second blow. Moose numbers in the 1930s dropped from an estimated two thousand to four or five hundred.

The winter of 1995–96 was more severe than any in over a century on the island. Lack of forage for the moose (i.e., too many moose eating too few balsam fir—see photos), a heavy outbreak of moose ticks, and the severe winter (very deep snow and temperatures regularly at or below negative twenty degrees Fahrenheit) all conspired against the moose. Finding food in deep snow was difficult and the depth of snow and cold temperatures meant the winter persisted longer than usual, resulting in a late spring, delaying spring growth which starving moose desperately needed to survive. Moose starved to death by the hundreds. The moose population collapsed from its all-time high of 2,400 to just 500.

Summarizing the math: An estimated 1,900 moose died in the winter/spring of 1995–96 with eleven or twelve dead moose per square mile of the moose-functional territory. At over six hundred pounds per moose, there would be more than 3.5 tons of putrescent moose per square mile rotting

Balsam fir nibbled to shreds. SOURCE: WOLF-MOOSE ANNUAL REPORT, 2021–22, KY KOITZSCH AND LISA OSBORN

away that spring. Yikes! I can testify that rotting moose is apparent from hundreds of yards away. There is no mistaking the smell.

This slow motion, smelly event in 1996 and the 1930s was likely on the minds of the Park Service as the decision to introduce new wolves to the island in 2018 and 2019 was being considered. Everyone knew what would happen to the moose population without predation. In fact, from 2012 to 2019, when wolf counts were very low, the moose population increased from 750 to 2,060 (estimates from the Winter Studies done in January/February of each year). It was a nearly three-fold increase in just seven years. Everyone knew how this moose abundance story would end.

What did the volunteers who look for dead moose do in 1996? I was not part of the project at that point, but can only imagine the messiness. The volunteer program in 1996 was relatively new, having only started in 1988. All off-trail hiking would have been full pack all the time, since GPS systems were not in use yet. The first volunteer groups for the year would have had to contend with snow and ice. The boats that brought them to the island would not be able to make it to the docks and volunteers would be let off on the ice in Washington Harbor at the west end of the island.

The volunteers would have found more moose than usual, and those they found, more often than not, would have died of starvation. There would be a lot of butchering done to examine bones of interest (i.e., skull and metatarsal at minimum) and carrying out a lot of meaty bones. Any volunteers that returned in subsequent years really had to love the project and dead moose.

CHAPTER 27

WOLVES, RADIO COLLARS, AND THE KINDNESS OF STRANGERS

IN AUGUST 2014, THE NUMBER OF WOLVES ON THE ISLAND WAS DWINDLING. There had been an estimated nine wolves during the Winter Study of 2014, mostly in one pack with some singletons around the island.

By August of 2014, the Wolf-Moose Project scientists had gone the entire summer without any sightings of the main wolf pack. The scientists, the Park Service, and the vacationers did not know where the wolves were. Were they on the east end of the island, the west end, the middle, or moving all over?

I had extra vacation time lying around, so I told Rolf that I'd come to the island to hike and try to find the wolf pack. One last wolf on the island still had a functioning radio collar, so I would hike the island with a telemetry unit to find that wolf, and if I found it, I would have found the whole wolf pack.

This would be a different sort of hike on the island since I would primarily be on trails the entire time as opposed to off-trail hiking, which is typical for the volunteers when looking for dead moose. The idea was to hike from one end of the island to the other, down short side trails with a few jaunts off trail to promontories, stopping all over and using the radio telemetry to try to find the wolves.

My travel route took me to Copper Harbor, then across Lake Superior via the *Isle Royale Queen IV*, and I landed at the east end of the island at Rock Harbor. I quickly got on the *Voyageur* for another boat trip of several hours to get to the western end of the island. The plan was to hike from Windigo back to the eastern end of the island. By the time I got off the *Voyageur* at Windigo in Washington Harbor, I'd had enough boats for a while. Diesel fumes get old quickly.

Telemetry units are notoriously imprecise, at least they were in 2014. The units we use on Isle Royale look like the letter H being held up in the air while you have an earphone in one ear and slowly rotate in a circle trying to pick up a signal. The stronger the signal is, the closer the radio collar and animal, and the narrower the angle of the signal, the closer the radio collar is. So if the signal is weak, it can be heard through a wide angle where the signal is "visible." You might be able to swing through

Scott Larson operating a telemetry in 2010 while David Norris looks on. SOURCE: JEFFREY MORRISON

ninety degrees and hear a weak signal. Or, if the wolf is close by, it might be a tight ten-degree wedge with a loud signal. One time when I had such a tight signal, I figured the wolf was very close and *watching* me even though I couldn't see it.

The point of the radio telemetry wasn't to locate the precise location of the wolves, but rather a range where they were—a vague area on the east or west side of the island. My plan was to find the wolves with a signal and then hike like a madman to another location—using trails—to "triangulate" their location. That was my plan, anyway.

At Windigo, after leaving the boat, I telemetried (now a *verb*). No wolves. So I continued hiking. It was early afternoon and I didn't feel like doing a lot of miles, so I left most of my gear at Windigo and went due south to Feldtmann Lake where there was a camp site. I wasn't going to camp there, but it was a logical spot on the map to try the telemetry again. I would get four or five miles in, telemetry (verb), return to Windigo, and do serious hiking the next day.

Hike to Feldtmann Lake. Telemetry. No wolves. Back to Windigo. Put up tent. Sleep.

When I got up the next day—my second day on the island—there was glorious sun and the temperature was great for hiking: chilly, but not cold. You start off hiking and wonder if you should have another layer of clothing on, but know you won't need it as you quickly warm up.

I needed to figure out where I was going. But not before telemetrying (no wolves). I started hiking to the east, straight down the Greenstone Ridge Trail from Windigo. This was the main and busiest trail connecting the west and east ends of the island. I decided I would hike around the southwest portion of the island the opposite direction from what I done the previous day. On this first full day, I would hike the Greenstone, hang a right, go south on the Island Mine Trail, past Island Mine Campground to Siskiwit Bay Campground, hang another right, and go west on the Feldtmann Lake Trail until I hit the fire tower. The hike would describe two-thirds of a circle going clockwise.

This route covered ground I hadn't covered the previous day. It was an inefficient way to get to the fire tower, which was roughly twelve miles away. If I had simply taken the trail counterclockwise, I would have hit the tower in about ten miles. As I write this years later, I am not entirely certain why I decided on this route other than thinking that I might have picked up a signal from the Greenstone Ridge Trail, making the trek to the fire tower superfluous.

Naturally, I didn't get a signal from the Greenstone Ridge Trail. I continued around the circle in a clockwise fashion. It was a beautiful, sunny, and—eventually—warm day. I hiked quickly given that I had a full pack—forty-five pounds to start—and a fair number of miles to cover. This hike had a totally different goal than I'd ever had on the island before and I was semi-cluelessly enjoying it.

I telemetried several miles outside of Windigo on the Greenstone, then someplace on a promontory with an opening in the canopy, then along the Island Mine Trail, then at Siskiwit Bay Campground. By this time, it was starting to get really hot. But there was no wolf collar signal.

I ate lunch at Siskiwit Bay Campground and noted there were several open campsites and shelters. I had the thought to go to the fire tower and come back to the Siskiwit Bay Campground and camp. So I stashed my pack just outside of the campground, GPSed its location on the off chance I completely forgot where I stashed it, and I hiked to the fire tower sans full pack.

At this point in the day, I was getting tired. But hiking without a full pack was nice. I got to the fire tower—a rickety old thing. Rolf had given me the keys to the tower so I could access the balcony and look around more easily. If the wolves were on the western end of the island, I figured I ought to be able to pick up a signal from the fire tower. I climbed the tower, unlocked it, got to the balcony, telemetried, and . . .

Hot damn—a signal! But where, though? It was not a strong signal, meaning I got a distinct but squishy-soft return through forty-five-ish degrees north by northwest around a bearing of 350. Good information, but not *super* clean, definitive information. I wanted a signal that allowed me to tell Rolf that the wolves, at least when I found them, were at such-and-such a location and be able to be put my finger on the map at a fairly specific spot and not just a direction from the fire tower.

I stalled and ate a leisurely snack on the fire tower, enjoyed the vistas, and waited for nearly an hour. Then I telemetried again. The signal had shifted; the wolves were now north and a little *east* with a bearing of twenty. The signal strength was similar with a return through about forty-five degrees; distinct, but not strong. The wolves were probably about the same distance from me as they had been earlier, just moving due east.

What to do next? Stay and take more readings, or go? I conceived reading from another location to triangulate the wolves' location. I'd be ever so smart if it worked. And I didn't want to stay at the fire tower too long since I did not have my gear with me.

I decided to go back to the Siskiwit Bay Campground, get my pack, and telemetry again. I climbed down from the fire tower hiked back to Siskiwit Bay Campground. When I got there, the campground was full. I had just been there a couple hours before and now all the tent sites and shelters were being used. I considered cramming my way into a campsite and telling whoever was there to suck it up and get used to me being there. But I didn't.

I completed another telemetry reading and the signal seemed similar; again easily heard but not strong, and the direction, while perhaps more north-ish with a little less east-ish, suggesting the wolves might have stopped while I had hiked east from the tower. Or were they moving? I couldn't be sure. I considered more hiking for another go at sloppy triangulation.

It was still sunny and hot. Even so, I decided to go north about four more miles to the Island Mine Campground. This would be a long day of hiking, about nineteen miles, but it would make the following days easier. And I could do one more telemetry reading from roughly the direction the telemetry unit suggested the wolves were located. I saddled up and went.

I hiked slowly to Island Lake. Only one other group was there, a father with a son and daughter. The kids were down at a little stream playing as I hiked into camp. They were full of energy, the little scamps. The father was putting together a fire in the fire ring; Island Lake Campground was one of the few spots on the island where you could have a fire. I set up my tent and got my stuff out.

I took a last telemetry reading for the day, and the wolves seemed to have moved. They were due west from my current location. I had hiked north or toward where the earlier signal suggested they would be, and now they had moved. How inconvenient. Wolves move around all the time, no surprise, but they were definitely at the western end of the island and moving around in the big empty area south of the Greenstone Ridge Trail and north of the Feldtmann Lake Trail—an area where nobody hiked much since there were few trails and it was mostly swamp.

Island Lake was a funny campground in that it was totally buried in trees, had no views, and no vistas. But it was a logical location to crash at the end of a long day of hiking. I visited the father and kids at their fire and demonstrated the telemetry unit, thinking I'd be able to let the kids listen in and hear the signal beeping away and they would get a nice little science lesson. Except of course there was no signal when I demonstrated this—making the telemetry unit seem boring with only soft static. I think the father was politely interested in what I was doing, but the kids didn't seem impressed. The wolves were obviously on the move someplace.

I went back to my tent to crash, but not before I learned there was a large series of thunderstorms coming to the island in three days and would last for several more. In 2014, I was still carrying a little "transistor" radio on the island so I could listen to the weather from one of several distant, scratchy-sounding radio stations. The storms sounded like something I'd want to miss if possible. Sometimes storms on Lake Superior were intense enough where the boats didn't run for a day, and I didn't want to get stuck on the island for an extra day. At this point in my hike, it was a challenge since I had a lot of miles to cover before I was planning on leaving.

I fell asleep and slept soundly. I got up early, ate a quick breakfast, and broke camp before the father and kids were up. And—of course—telemetried before hiking (no signal). Back up to the Greenstone Ridge Trail to hike, hike, hike.

I spent the day hiking the Greenstone Ridge Trail and occasionally stopped to check for wolves. I passed Mount Desor, and at the little feeder trail that goes to the South Lake Desor camp, I bumped into a group of six guys from Iowa who were fishing and hiking and camping. They were also hiking to the east. I chatted with them briefly and then I went ahead of them on the trail.

Shortly after talking to the Iowa guys, I stopped, stashed, and GPSed my backpack, and went off trail to a *great* promontory along the Greenstone Ridge. I bushwhacked to the spot and found a lovely flat shelf overlooking a large stretch of the island toward Lake Siskiwit to the south. It was covered by a wild lawn with flowers. It was gorgeous. It was perhaps the prettiest spot on the island I had ever been.

I enjoyed a long leisurely lunch, watched bees on the flowers, and saw an eagle in the distance flying. I took another telemetry reading (nope, nothing), and ate a bit more. After about a half hour, I decided to leave since I had more hiking to do.

I returned to the trail and hiked on, eventually arriving at the Ishpeming Fire Tower. The fire tower wasn't much of a tower since many trees in the area were taller, making visibility from the tower nil. The Iowa group was there, resting and eating a snack. I hiked up to them and they looked perplexed since I had been ahead of them on the trail the last they knew, and now I was coming up behind them. I stopped and ate a snack and explained what I was doing. I learned they were heading to Hatchet Lake that afternoon, which is where I planned to stop for the day.

I finished my snack and off I went again in front of the Iowa group. Between the fire tower and Hatchet Lake was around five miles or so. I went off trail once or twice to telemetry from more promising spots.

When I got to Hatchet Lake, it was getting late and I was tired. And there were no open campsites. I found the Iowa hikers at a group site, which had some open spots. I bulled on in and informed them I was pitching my tent. They were fine with this.

Through the evening, I sat with them for a while talking, but eventually I went to bed since I was tired and I had notes to finish up. I had not gotten any wolf collar signal since the previous day. I presumed the wolves were behind me and I wouldn't pick up their signal again.

Before sleeping, I once again listened to the little transistor radio and learned the impending storm was set to hit two days out *and* to last for several days. *Ugh.* After hearing that, I decided to complete the hike from Hatchet Lake to Daisy Farm the next day—this would be over sixteen miles of hiking. Sixteen miles with a full pack was a chunky number of miles, especially since I'd already hiked about fifty miles in the previous three days. But my preference was to finish the hike before the storm hit, especially since I had located the wolves and they were roaming the west end of the island, not the east, which was the direction I was hiking in. The Iowa guys were still talking when I fell asleep.

I was up early the next morning and gone and hiking before the Iowa group woke up. There was a long hike ahead of me and I didn't lollygag. I went straight down the Greenstone Ridge past West Chickenbone and then

East Chickenbone. It was a warm day. Every so often I stopped for snacks, water, and to take telemetry readings. No signal, no surprise. I took the Daisy Farm Trail and cut down to Moskey Basin and Daisy Farm.

I arrived at Daisy Farm around 4 p.m. The sun was out, Daisy Farm was fairly packed with people, and I had no easy way to communicate with Rolf at Bangsund Cabin. Bangsund Cabin is across Moskey Basin and perhaps one-third of a mile of open water. And I was a day earlier than I had told Rolf I'd be here.

There was an unattended boat at the Daisy Farm dock. I asked everyone on the dock if it was their boat, but nobody owned up to it. So I plopped my backpack down next to the boat and settled in to wait until somebody showed up. I figured I would "hitch" a ride across Moskey Basin to get to Bangsund Cabin—I'd do my best Blanche DuBois and depend upon the kindness of strangers. One of the lovely things about backpacking and being in the backcountry was that being presumptuous and imposing yourself on others—I try to avoid this, of course—was almost always handled with grace and willingness to help out.

Not too much later, two brothers, Robert and James, both in their mid-sixties and from someplace out east, showed up. They had been hiking up to the Mount Ojibway fire tower and back and were now getting ready to motor back to Rock Harbor and the Lodge for the night. They had also heard about the approaching storm and were leaving the next day on the boat—the boat I planned to be on.

They motored me over to Bangsund Cabin as requested. *Thank you very much.* The brothers were introduced to Rolf and Candy and got a quick tour of Bangsund and the world's largest collection of antlered moose skulls. Then we said our good-byes, I told them I'd see them the next day at Rock Harbor and on the boat, and they were off.

I told Rolf what I had learned about the wolves, then spent what was left of the day relaxing. I felt I'd earned a rest since I'd hiked over sixty-eight miles in not quite four full days. And my feet hurt. In fact, I had bled from *around* my toenails, which I had never experienced before or since. No blisters though, so that was a win.

The next morning, Rolf motored me to Rock Harbor to catch the boat where I got on just as the weather was starting to turn. It was gloomy, cloudy, and rainy.

On the boat, I sat and talked with Robert and James. They let me know they were leaving the island a day early because of the weather too. They had originally flown to the island and, because the plane was full today, opted for the boat back to the mainland. This prompted the topic of how to get from Copper Harbor, where the boat would land, to the Houghton airport, where their car was parked. Having never been to Copper Harbor, they assumed they could simply get a cab. I told them a cab probably wasn't an option, and it was unlikely Uber would be an option either. I told them the nearest taxi might be in Houghton/Hancock or perhaps Calumet, each of which was far enough away that I doubted anyone would pick them up.

Since I depended upon the kindness of strangers the previous day, I offered to drive them to the airport since my car *was* in Copper Harbor. Problem solved. We landed in Copper Harbor just before 6 p.m. and drove the hour or so past the airport in the rain. From there we drove into Houghton and checked in to our hotel rooms.

We agreed to meet in Houghton for dinner. We did, and had a lovely dinner with an enjoyable time talking. All because of chance and the kindness of strangers meeting on the island.

CHAPTER 28

WOLVES AND GENETIC RESCUE 2

By 2015, there were only three wolves remaining on Isle Royale. The impact of the Old Gray Guy and his infusion of new genes had long since dissipated. What happened?

For a number of years, a combination of inbreeding, a skewed sex ratio with not enough females, and the death of three wolves in the mineshaft during the winter of 2011–12 caused the wolves' numbers to dwindle to just two adult wolves, identified as M183 and F193, and their nine-month-old pup (the last of the island-born wolves). These were the last wolves from the Chippewa Harbor Pack. The photo shows the pup on the far left, M183 (male) in the middle and F193 (female) on the right in 2016. The Old Gray Guy was M183's great-grandfather. The pup had a visibly abnormal tail, and would have been a highly inbred animal.

The loss of the three wolves in the mineshaft was especially critical as an event that underscored the precariousness of the wolf population on Isle Royale. Mines were part of the island's nineteenth-century legacy; most mine shafts were basically ragged holes dug straight down into the earth—think "pit." In the winter, mines—filled with water—had a skin of

The last three native-born wolves in 2016. SOURCE: WOLF-MOOSE PROJECT, ROLF PETERSON

ice. It was thought the wolves entered the mine to investigate something, and the combined weight of three wolves was too much—the ice broke, and they couldn't escape the mine. They all died. Most of an entire wolf pack extinguished along with a lot of the DNA variability available on the island.

The three wolves included the pack's alpha male, another radio-collared male—which is how the researchers found the dead wolves—and a young pup. Likely, M183, age two at the time and part of the Chippewa Harbor Pack, would have been witness to the mine shaft disaster. F193 was a pup, roughly nine months old, and might have been present too.

This event left the island with only nine wolves scattered into one pack of six wolves, a pair of wolves, and a singleton wolf. Singleton wolves are wolves that have left their pack and would like to start their own pack. At the time, this was the lowest number of wolves ever recorded on the island. Worse still, this count of nine only included three females, two of whom were older. F193 was the youngest female on the island.

Wolves are rarely radio collared, so when, where, and how a wolf dies is not usually known. The count of nine wolves in 2011 continued to drop. Fast-forward to 2016, and the island only had the two remaining wolves: M183 and F193. DNA analysis confirmed that this last male-female pair both originated in the Chippewa Harbor Pack, born to the same mother. F193 was also the daughter of the male, M183; they were simultaneously father-daughter and siblings. Their chance of successful reproduction was low.

Also during this time, the scientists spent several years calling for an infusion of new wolves from the mainland to the island as a genetic rescue. However, there was a large problem. Isle Royale was a wilderness area. "Wilderness" in this context is not simply a generic noun, but also a legal status covered by the Wilderness Act of 1964. The Wilderness Act aimed to preserve an

> area where the earth and its community of life are untrammeled by man, where man himself is a visitor who does not remain. An area . . . of undeveloped Federal land retaining its primeval character and influence, without permanent improvements or human habitation, which is protected and managed . . . to preserve its natural conditions.

Over the years, the National Park Service and other federal entities had a semi-official hands-off policy with no interventions for wilderness lands. Bottom line—the Park Service did not want to introduce new wolves, they wanted to let nature takes its course; maybe the wolves would make a comeback. The scientists pointed out that humans had already "intervened," if indirectly and unintentionally. Due to humans and climate change, the wolves had to contend with less frequent ice bridges, a mine shaft killing off 25 percent of the island's wolves, and parvovirus. The debate on whether or not to introduce new wolves stretched over several years.

Genetic Rescue and New Genes

After several years of consulting with numerous scientists and policy makers, and with public feedback, the Park Service determined they would introduce new wolves to the island. Before making this decision, the Park Service considered four separate options:

1. Do nothing—let nature takes its course.
2. Immediate limited introduction—Introduce one or more new wolves in a five-year time period.
3. Immediate introduction, with potential supplemental introductions—Basically, do option #2 with the possibility of more new wolves—if needed—later.

4. Wait and see—Essentially do nothing, but monitor and consider options in the future.

Unfortunately, by the time the Park Service decision was made—officially option #2—the opportunity for "generic rescue" had been missed. Genetic rescue was a mitigation strategy to restore genetic diversity within an inbred and isolated population, ideally to reduce the risk of extinction of the wolves on the island. However, by the time the new wolves were introduced, M183 and F193 were old for wolves in the wild.

Regardless, in late 2018, the Park Service began capturing and releasing new wolves on the island. In all, nineteen wolves, all with GPS collars, from four mainland locations were relocated to Isle Royale. The new wolves were larger and heavier than their predecessors. Females in the Great Lakes region average fifty to seventy pounds while males are sixty to eighty pounds. The goal of the introduction was to increase the numbers of wolves and to restore a healthier predator-prey relationship on Isle Royale.

The wolves came from several mainland locations and involved coordination with a number of entities. Four wolves came from Minnesota in cooperation with the Grand Portage Indian Reservation. These were the first wolves to be relocated in 2018. Eleven wolves were relocated from Ontario, including eight from Michipicoten Island, a large island in eastern Lake Superior. And three wolves came from the Jostle Lake, Ontario, area, northeast of Lake Superior. The wolves from Michipicoten Island were relocated, in part, because of the concern they had essentially eaten most of the prey on Michipicoten Island. The last four wolves came from the Keweenaw Peninsula area in the Upper Peninsula of Michigan in late 2019.

This photo is from the National Park Service and shows a Minnesota female wolf being released on Isle Royale, September 26, 2019. You can see her collar and her ears are tagged. She's beautiful and obviously frightened.

Both F193 and M183—the last two native-born wolves on the island—died in 2019. F193 disappeared sometime during 2019, never to be found. On October 17, 2019, as Isle Royale was shutting down for the year, Michael Ausema, a park ranger, was completing his last patrol on the Hatchet Lake Trail near the middle of Isle Royale, when he came upon

A female wolf being released on Isle Royale in 2019. SOURCE: NPS WEBSITE, JACOB W. FRANK

M183 dead on trail, October 2019. SOURCE: WOLF-MOOSE FACEBOOK, ROLF PETERSON

the body of M183, the park's last male island-born wolf. The dead wolf was lying on his side on the trail, looking almost like he had stretched out for a nap. The fact that M183 died in a very visible spot on that trail—and that somebody happened to be hiking there just a day or two later—was an incredible gift to the scientists.

An autopsy showed that M183 was killed by new wolves that were introduced by the Park Service in 2018–19. F193 was also likely killed by the new wolves. The genes of M183 and F193 disappeared from the island.

The introduction of new wolves was too late for a genetic rescue, but it has been a great success in terms of re-establishing the wolf population on the island. In the first three years on the island, at least five litters of pups were born. One litter of pups was born in 2019, two litters in May 2020, and two more in May 2021.

The 2022 Winter Study determined there were twenty-eight wolves on the island; this was double the count from 2020. The 2021 Winter Study did not occur—COVID got in the way. As of the 2022 Winter Study, the wolves appeared to be living in two packs: one in the island's western region and the other in the eastern, each with at least thirteen wolves and two wolves that may or may not be part of either pack. Evidence indicates the reintroduced wolf population is well established and healthy.

Since the reintroduction of wolves, every Winter Study, the scientists are very interested in determining how many pups were born in the prior year. What will the wolf count could be? As high as twenty-five, thirty, or forty?

And, due in part to the increased wolf predation, the moose count is going down. While the Park Service didn't stress this much in their decision to introduce new wolves, reduced moose count and the impact of large numbers of moose on the island ecology was a concern. Too many moose eat too much, then there is not enough to eat, and at some point, a lot of moose starve to death in the late winter and early spring, making a large mess around the island come springtime. Which is exactly what occurred in 1996.

CHAPTER 29

EVERYTHING IS CONNECTED TO EVERYTHING ELSE

SOME DAYS YOU HIKE AND ARE GOAL ORIENTED WITH SPECIFIC COORDINATES or locations to visit. You aren't hiking slowly or deliberately in picket lines since you don't have the intent to carefully search a swath of territory. When you are on a goal-oriented hike, it is often easier to simply jump onto a trail and quickly zoom from one place to the next. You use the trails for a while, then veer off into the woods, find your coordinates and a dead moose, do the off-trail backcountry CSI thing, and then get back on the trail to head to the next location of interest.

You don't find dead moose, or much of anything of scientific interest, on trail, of course, but you get to your destination more quickly. Jumping onto a trail is easier on the east end of the island where there are simply more trails where you can avoid ponds, lakes, and swamps and crawl over ridgelines more easily. Perhaps you might not totally avoid obstacles, but they are easier to get around.

One year my group and I were doing exactly this—taking a trail to get from point A to point B more efficiently. We had spent a portion of the week hiking and exploring an area north of Tobin Harbor and south of the Greenstone Ridge. We started the day way off trail and were finishing up by hiking into Three Mile Campground using a short connector trail from the Greenstone Ridge to the shore. It was later in the week and I was planning on spending two nights in Three Mile Campground and doing day hikes from there to end the week.

The area we had just left was prime moose territory with plenty of food, water shield, aquatic plants, balsam fir, and so on. Moose territory meant *wolf* territory, but this hike occurred in 2018 and the wolf population was waning and moving toward extinction on the island.

My group stumbled out of the backcountry onto the trail. This trail ran mostly north-south for several miles and straight into Three Mile Campground. Three Mile is so named since it is three miles from Rock Harbor, the main landing point for everyone on the eastern end of the island.

After being off trail with a full pack, hitting a trail is a relief. You don't need to pay as much attention to where you place your feet, you don't need to look down constantly trying to discern moose bones from the detritus on the ground (aspen branches fool everyone), and I, as group leader, do not need to keep track of everyone, hovering like a mother hen. It is a nice break from off-trail bushwhacking.

We hiked along all happy and such. It was almost the end of the day as we hiked to a campground where there would be level tent pads, picnic tables, and we'd be on water—chilly Lake Superior.

Soon we were hiking next to some water. On the right was a large pond of perhaps four to six acres. It was a pond that should not have been there. As a group leader I—naturally—do not want to get lost—at least not too often or too badly. I have to inspire confidence in the volunteers. On occasion I will get a bit turned around for a minute or two, but I do not get well and truly lost.

The pond confused me. I had been on this exact section of trail the year before, as well as many other times, and there shouldn't be a pond where there very clearly *was* a pond. Were we where I thought we were? I stopped the group for a break. I checked my compass. Yes, we were heading and facing south, which was what we should be doing. I already *knew* that because the sun was out and we were facing it. Even so, I took a quick GPS reading and cross-referenced it with my map. Yes, we were exactly where I thought we were. And the map didn't show any pond. I was still confused, but now confident we were where we were supposed to be.

The pond was brand-spanking-new since I'd been there the year before. I scanned the pond and finally found a big whomping beaver dam at the southern edge of the pond with water spilling over the trail here and there.

This was where the trophic cascade concept comes in. We were in primo moose—hence, wolf—territory. A beaver should not have just waltzed in without the wolves' by-your-leave and decided to build a dam and create a large pond. A healthy wolf population discouraged this type of thing.

Trophic cascade is a domino effect that hits environments. "For want of a nail the shoe was lost, for want of a shoe the horse was lost, for want of a horse." The littlest thing can and does matter. Everything is connected. And sometimes, one of those critical dominos falls, one that impacts everything, quickly, and is noticeable even to amateurs like me.

The cascade was pretty simple in this case. Wolves were dying out on the island because of fewer and fewer ice bridges to the mainland and Canada. When the ice bridges *did* occur, they were thinner and shorter lived than previously and more susceptible to the wind and temperature fluctuations. The ice bridges, when they formed, broke up quickly, and no wolves could cross. Climate change impacted a phenomenon that—before it occurred—you might not even have considered.

Because no wolves crossed from the mainland *for years and years*, the wolf population became increasingly inbred. The wolves' presence on the island was precarious and it wouldn't take much to tip them toward disaster. The big domino that fell—*loud thud*—was in the winter of 2011–12, when the three wolves from the same wolf pack went into the mine shaft and died.

After the remaining wolf count continued to fall, wolf predation, not just of moose, but *all* animals, dropped. Fell off the cliff. And every other species on the island noticed. Beavers thrived: they moved all over the island from isolated little pockets to essentially every bit of moving water from streams, to trickles, to seeping springs. Beavers had an opening, and they took it. They built dams everywhere, including the lovely new pond we were hiking by.

There are subtle signs of change I've noticed on the island from year to year, such as changing water levels, especially in swamps; pitcher plant numbers expanding or contracting; and balsam firs being eaten or totally tortured by the moose. But the beaver dams and resulting ponds are super easy to see. And this was not the island's only new beaver dam due to the lack of wolves.

My group and I finished our water break and went on our way. As we passed the southern edge of the beaver dam, I looked it over critically; off-trail hikers on Isle Royale pay attention to beaver dams. It was new. Sticks on the dam were freshly chewed and some had sap running out of them. There wasn't much, if any, growing grasses in the middle of the dam. The lodge was obviously in use with new mud dug up here and there. Tree trails where the beaver had dragged branches to the pond were fresh. This beaver dam was a going concern.

We continued hiking. Got to Three Mile Campground. Closed out our day in a campground, for once.

In 2019, new wolves hit the island, re-introduced by the Park Service. The dominos that fell in 2011 and other years were picked up and put back in place again. The moose and the beaver noticed. I haven't hiked that section of trail since the reintroduction of wolves, but I'm looking forward to retracing my steps to see if the beaver dam is still there. And still active.

CHAPTER 30

WHAT WE EAT IN THE BACKCOUNTRY

FOOD AND EATING WHILE BACKPACKING ANYWHERE, NOT JUST ISLE ROYALE, CAN be complicated. While hiking, you want food that is easy to make, easy to clean up, is lightweight, doesn't take up a lot of space, isn't super expensive, has some variety, is shelf stable (i.e., doesn't need to be refrigerated), and—ideally—it should be delicious. And when you make food, you want to make *exactly* the right amount. Too little, and you are still hungry. Too much, and what do you do with it? You cannot put it in the Tupperware and slip it in the refrigerator, and you cannot just throw it out. There are Leave No Trace rules about that kind of thing.

On Isle Royale, volunteers bring their own food. I've seen quite a few miscalculations, both in terms of quantity but also missing on one of the other "facets" (e.g., it's hard to make or tastes awful). The longer a backpacking trip is, the more important each of these factors becomes. For instance, if you miscount the amount of food and you have too much for one or two days, it isn't such a big deal. But if you are carrying an extra pound of food per day for seven or eight days, that is a lot of extra weight to carry around for the week. And no, sharing with others generally doesn't work because they all have more than enough food, too, and are trying to share with you. People tend to bring too much food when they are packing food for just themselves.

Breakfast food is fairly standard with many people simply bringing instant oatmeal (which I abhor) or sometimes energy bars (more my style), and I always have coffee—real coffee, not instant. Breakfast is also sometimes a short time period especially if you are tearing down tents and camp in preparation for a full-pack day of hiking.

Lunches and snacks tend to be things like nuts, gorp (trail mix consisting of nuts, chocolate, fruits, and so on), chocolate bars, beef jerky, cheese, crackers (which are crumbs by the end of the week), peanut butter (which is heavy), dehydrated fruits, and so on. There is nothing strange here; it is mostly snacky food stuff you would find in your average grocery store.

Dinners are where things get complicated. After a long day of eight to ten hours of hiking off trail, most people do not have a lot of energy. The thought of having five or six of us try to coordinate making dinner and cleaning up with one or two camp stoves sounds involved and clumsy. For my groups, I always prepare camp dinners for all of us before we hit the island. I've been doing this for years and everyone seems to be pleased about it since making dinner is much more efficient time-wise.

Before we get to Isle Royale, I pester Rolf to let me know who will be in my group so I can contact them to get any dietary restrictions and preferences. Over the years, I've developed a fairly lengthy list of entrées that are quite tasty and easy to make. This includes burritos (bean and/or meat); chicken, rice, and broccoli; chicken pesto pasta with parmesan; chicken curry with peas; beans, rice, and ham; lemon-dill rice with salmon and broccoli; chicken gravy train over mashed potatoes; and a number of soups with side dishes (e.g., couscous or polenta with additives like fresh vegetables). All of these dishes can be modified to be vegetarian and/or gluten free.

I scale them to *ideally* make exactly enough food for each of us; I err on perhaps a slight underestimation since it is easier to add other foods than it is to handle too much food. The extra food for my dinners are desserts and hot chocolate. The desserts (e.g., biscotti and Rice Krispie Treats) are individually wrapped so people can have one or more, or skip them entirely if they've had enough to eat.

Raw ingredients for spaghetti dinner for six. SOURCE: MEGAN HORODKO

One dinner I make for nearly every packing trip is pasta with sauce, parmesan cheese, and dry Italian salami. I'll use this dish as an example of what is involved in a single meal. At home, I prepare the component parts and package them up. The photo shows the individual ingredients to make a dinner for six.

- For backpacking, I use rotini or some kind of pasta that is easy to eat with a spoon. Backpackers usually eat out of a bowl or cup and rarely have forks or knives so regular spaghetti would be clumsy to carry, to cook, and then to eat. A shorter cooking time is better to conserve stove fuel. I typically bring three ounces of pasta per person.

- Spaghetti sauce is the most involved part. I generally get around thirty Roma tomatoes, dehydrate them, and then use the food processor to create a powder.
- I add a couple of McCormick Thick and Zesty packages to the tomato powder, which has a thickener, corn starch, and spices, though any off-the-shelf spaghetti sauce additive would probably work. Then I add more spices like basil and oregano. The tube in the photo is double-strength tomato paste that also acts as a thickener as well as being tasty. The tomato paste also seals nicely if it is not all needed for the sauce and can be used in other dinners.
- The parmesan is a real mid-priced cheese, and there is a tiny grater.
- Last is the Italian dry salami.

Preparing dinner next to Lake Shesheeb. SOURCE: ALEC SMITH

Voilà, dinner! SOURCE: MEGAN HORODKO

I package all of these items up in a gallon Ziploc bag and include a little slip of paper with preparation instructions. The total weight for six people is about three pounds, or eight ounces, per person.

This is a two-pot meal: one pot for simmering the sauce and the other for making the pasta. Some of this is a little tough to see in the "prepping dinner" photo. Cecilia is grating the parmesan into the plate in front of her. The dry salami is waiting for someone to slice into the other plate. I have two cook stoves and two pots in front of me ready to go, with my hands cupped waiting for someone to throw me a lighter (lower right of photo).

It all gets cooked up, everyone assembles their dish, and then we eat. This particular dish ends up being very tasty, and every bit as good as homemade. After we eat, we boil up another pot or two of water for hot chocolate—I dip biscotti in my hot chocolate—as well as to clean dishes.

As with most dinners, me and one or two of the volunteers help prepare and clean up, all with less fuss than if we were all doing our own thing. A pleasing communal meal is how I like to end a day of bushwhacking.

CHAPTER 31

WOLFIE COMES A-CALLING

IT WAS JUNE 2023, AND IT HAD BEEN A LONG DAY. WE HAD GOTTEN UP EARLY to catch the boat to the island, landed at Rock Harbor, then took a twenty-minute boat trip west down Moskey Basin to Bangsund Cabin, where we got organized, ensured we had our complete gear, ate a little bit, and Rolf let us group leaders know what we'd be doing for the week. By this point, it was after 3 p.m. *Then* Rolf took my group in the boat again down to the western end of Moskey Basin to the Moskey Campground, where we started a six-mile hike west to Lake Richie, then meandered north around Lake LeSage and Lake Livermore, crossed the Greenstone Ridge, and finally landed in West Chickenbone Campground. *Whew.* Long day.

At West Chickenbone, we set up camp at a group site that was about forty yards from the water. We then made our dinner. It had been a pretty day and the evening was also lovely. Four of us went to the edge of the lake and watched the evening sunset and talked quietly so we didn't disturb the beaver that was in the water swimming back and forth in front of us.

That was when two of our group, Cecilia and Jacob, came up to us hurriedly, obviously geeked. They had seen a wolf.

Jacob had been in the outhouse and Cecilia was waiting for him to finish his business when she turned around, and about twenty feet away on the trail was a wolf just watching her. She was *not* comfortable with this—was frightened, in fact—and knocked on the outhouse door and told Jacob to come out because there was a wolf. Jacob thought she was just kidding around, but when he was done with his business and came out, he saw the wolf. The wolf ran off. And then Jacob and Cecilia came to us, geeked. As they related all of this to us, Alec, another member in our group, pointed down the trail and said, "There it is!" We all leaned over and sighted along Alec's arm to see where he was pointing, and saw the wolf.

The wolf looked back at us, considering us, and then turned away. I barely figured out where Alec was pointing when the wolf disappeared. So while I *saw* a wolf, it wasn't a clean, long, lingering look. More of a sketchy glance.

A second or two after the wolf disappeared, I started back up the trail to our campsite where the wolf had been, since I wanted to see more of it. Every one of us explored around the various trails for a few minutes but did not see any more of the wolf. We did find wolf prints at several spots. We eventually gave up and returned to our campsite.

All we found were wolf prints. SOURCE: JACOB DEPPER

Back at our campsite, Anna asked where our water was. Our filtered water from the gravity filter—a 1.5-gallon bladder on the ground—had been dragged by the wolf off into the weeds about fifteen feet. Fortunately for us, the wolf had used the handle and not the plastic reservoir itself which would have punctured the bladder. Nothing else was amiss around our campsite.

Seeing a wolf was exciting, but also concerning. These newer wolves, the newly transplanted wolves since the collapse of the wolf population in the early 2010s, weren't as shy around humans as the wolves pre-collapse had been. Pre-collapse wolf sightings were rare and the wolves used to howl frequently. The newer wolves weren't as shy and there have been numerous sightings, and—much to my chagrin—they don't seem to howl as often. I love hearing them howl.

While I'd love to get a better look at a wolf someday, it really would be best if we humans only saw their tracks and scat and heard them howl.

CHAPTER 32

TYPE 2 FUN

NOT ALL FUN IS CREATED EQUAL. GROWING UP, MY PARENTS LOVED TO GO TO Florida every year. This was my family's main vacation and my parents loved it. It wasn't adventurous, but it was fun. It was Type 1 fun—something that you know will be fun as you are doing it and fun afterwards as you think about it. Also just known as plain-old "fun."

Having been to Florida multiple years as a kid, I always wanted to do something other than go to Florida. When I was nineteen, a friend and I hopped in my car (a rattletrap, rusty Datsun B210) with a couple hundred dollars, no cell phone or GPS (they wouldn't be invented for another fifteen or twenty years), a Shell gas credit card, and a lot of naiveté, and drove around the country for a month before we started college in September. We ended up sleeping in the car at the side of the highway several times. We got lost after hiking for hours in Zion National Park with no water in the ninety-degree heat. We were harassed by drunken patrons in a bar parking lot in the middle of Texas on a Saturday night, and so on. Much of the trip was about new stuff, uncomfortable moments, and some stupid what-were-we-thinking bits like almost falling several thousand feet into the Grand Canyon. It was an adventure where it wasn't always fun in the moment, but it was in retrospect. That is Type 2 fun.

Even now, after having hiked on Isle Royale for over twenty years, I don't believe my mother understands Type 2 fun; she thinks "vacations" should only be Type 1 fun. Okay—a lot of people think that way.

Most every backpacking trip will have some Type 2 fun. Off-trail hikes on Isle Royale for seven or eight days nearly always entails some Type 2 fun.

Weather

You're in it; you cannot escape it. Rain is probable and even snow is possible. My groups have hiked in snow—over our knees—in May and into June and you're almost guaranteed to be rained on at some point.

Rain on the island often comes at night, but in 2003, my group returned to camp from a day hike in the rain. The rain proceeded to get heavier. Somehow, we did not have a tarp and therefore, making dinner was going to be in a heavy downpour. I was not leading yet, so I opted to bail and went into my tent as the rain got harder and harder and the lightning and thunder more and more insistent. The rest of my group managed to get dinner prepared. I have no idea how they did that. I did not fall asleep until 11 p.m. and it was still a deluge; it was the hardest and most persistent storm I've ever been in before or since.

The next morning was, of course, glorious and sunny, but we all had a lot of wet gear. When I woke up, my tent was in the middle of a small pond about three inches deep. One other tent in our group was in the same pond. Fortunately, we had good tents with "bathtub" bottoms (i.e., waterproofing up the sides for six to nine inches). We were dry, but we had to carefully empty the tents, walk barefoot in the pond, and take our gear to higher ground. Once we had our gear on higher ground, we spread the tents out to dry.

This is not uncommon for backpacking. A lot of rain can mean wet gear. Wet gear means you find spots to drape it over to dry, even if you only have a short twenty-minute break.

This is Type 2 fun with a couple lessons: Tarps are useful, and maybe look for low spots when you pitch your tent.

What rest breaks look like when you have wet gear and it is not raining. SOURCE: JEFFREY MORRISON

Greenstone Ridge

When hiking, I am usually on the east end of Isle Royale, and most weeks I cannot avoid having to climb up and down the Greenstone Ridge several times. The Greenstone Ridge runs the length of the island and is five hundred feet higher than the elevation of Lake Superior. Many years we climb the Greenstone loaded with moose bones at the end of the week.

There aren't any "easy" spots to climb the north face of the Greenstone; there are no trails for miles and most of the slope is steep or even just vertical cliffs. And, because we're off trail, hiking east or west to find an easier spot is both time and energy consuming, but also "easier" is a relative term. Climbing the Greenstone is tough even without a pack at the "easy" spots.

I don't attempt to finesse the slope. With me leading, my groups just have to put their heads down and go. We head directly to the slope and start climbing. Sometimes you are crawling, grabbing every tree/bush/weed you can for stability. You will gain ten or fifteen feet, stop, catch your breath, and figure out where the next ten or fifteen feet will take you. Often people in the group take different routes and somebody inevitably ends up way ahead or behind everyone else. We're never so spread out we don't know where each person is at, but climbing the ridge is a sweaty, dirty, solitary effort.

More Type 2 fun, but I am not sure there are lessons to be learned, aside from knowing every unpleasant situation ends at some point. We all make it to the top of the ridge, rest, drink water, and have a snack. And, while we recover, we know the climb is behind us and feel a sense of accomplishment.

Smelly, Icky Bones

Most weeks there will be several dead moose that have some meat on the bones. By "meat" I mean something well past its "use-by" date. Adjectives like "rotting" and "putrescent" come to mind. If the moose has been dead longer, it might be more like a tough jerky. Sometimes there is butchering to be done that involves handling of the dead moose bones. Flies, maggots, and other gory bugs might be part of the mix. Or moose ticks.

In recent years, butchering isn't always done right then since we can GPS the moose and come back months later to work on the dead moose. Eventually, the moose should have less meat because of scavengers (e.g., ravens, fox, bugs, etc.). I still butcher moose when the location is difficult to get to.

Regretfully, I find my power of persuasion with smelly dead moose has limits and it turns out a lot of backpackers are squeamish. More often than not, I am the one who does the butchering. However, some of the time, there are people who are all in for the butchering. Have at it, but be careful with the knife or saw. And wash your hands as best as you can when you're done.

That is definitely Type 2 fun. The opportunity to butcher a meaty, putrescent moose doesn't come along too often in life. Best you should take advantage of the opportunity when it presents itself. Grit your teeth, breath through your mouth, and get to work. Once it is done, it wasn't so bad.

This really is the point of the volunteer week: Find dead moose and bring back important data to the scientists, no matter what temporary unpleasantness you need to experience.

> If you anticipate you will be miserable, you are going to be. . . . For [those] of us that are on an adventure . . . misery is only a moment in time.*

> They weren't great moments. . . . But I never had a bad moment, I enjoyed all of it.†

It's hard to have a real adventure without some Type 2 fun. And what is life if you don't have an adventure once in a while? Maybe not hiking off trail looking for dead moose, but doing something new, getting outside your comfort zone, and learning. Who knows, after a while, the Type 2 fun might become Type 1 fun.

---

* Amy Godwin, volunteer 2024, reflecting on hauling a rotting dead moose out of a pond in the cold rain.

† Jeremy Sartain, volunteer 2024, dead moose wrangler, in reference to the same dead moose, same pond, same cold rain.

CHAPTER 33

THE EMERGENCY WEEK

BEFORE HEADING OUT FOR A WEEK IN THE BACKCOUNTRY WITH MANY OFF-TRAIL miles in our future, every group can count on a farewell from Candy Peterson. Candy will tell us to have fun, to take care of each other, and she'll exhort us to not get hurt. Most weeks, this all takes care of itself. Injuries tend to be minor, like scratches, pokes with sticks, bug bites, and the like. Most weeks, but not *all* weeks.

My group for the summer of 2024 was unlucky, and rather than having a good week off trail looking for dead moose and seeing the sights, we instead were tested on our calmness during an emergency. Not once. Not twice. But three times. All in one of the hottest weeks I've spent on the island; it was at or above eighty-five degrees all week.

On our first full day, my group of four headed south from Lake Richie Campground where we had spent the previous night. Our main goal for the day was to stay on trail for a couple miles, then head off trail to the west for a three-mile bushwhack up and over a steep ridgeline around Chippewa Harbor to Blueberry Cove. If we could do that in one day, full pack, and find a good spot to camp near Blueberry Cove, the day would be a success. I was looking forward to the hike because I'd never been to the area around Blueberry Cove.

We followed the trail south and looked for a spot to head off trail to cross a little stream that ran between Lake Richie and Chippewa Harbor. For quite a way, the stream and trail shadowed each other. Crossing streams, or any kind of water, is always a bit dicey because most people don't want to get wet boots or—worse—fall into the water, especially with a full pack.

This particular stream was so wide at parts where we could not get over it easily, and so fast and narrow at others that the current looked threatening. We eventually found a spot that looked promising, perhaps fifty feet from the trail. It was narrow and looked as though we could get across the stream with some scrambling on rocks. I led the way, carefully avoiding wet moss, *especially* the patch of wet moss on an inclined tilted surface, and finally crossed with a large step and grabbed some rocks on the other side. Once across, I turned to let everyone know to be careful of the rocks with the moss.

Ze was behind me and immediately stepped squarely in the middle *that* patch of wet moss on the inclined rock. She went down hard. She slid into the fast-moving stream, her left foot wedged in between rocks, and she continued her "fall" downstream with an assist from the current. David, who was behind Ze, and I grabbed her to stop her from going further downstream. Ze struggled to get up, and between the three of us, we managed to get her out of the water and up onto dry rocks by the edge of the stream. Ze was obviously in a lot of pain.

We got her backpack off and gathered around her. After a minute or so, Ze took her boot off to see what the damage was. As the boot came off, Ze's leg came out. But her foot didn't immediately follow. I could see the leg and foot bones weren't connected like they ought to be. Ze passed out. We had an emergency.

After a minute, Ze woke up but was in pain. First, we had to get her away from the stream and ideally to the trail where there was room to help her. We made a brief, clumsy effort to stand her up and semi-carry her, but that didn't work with the bad footing by the stream. Ze crawled through the brush and downed branches to the trail. She was a very good crawler and was determined to get to the trail. Once on the trail, Ze said she could crawl to the dock at Chippewa Harbor, but seeing as Chippewa

Ze's splint—I love duct tape. SOURCE: BRIANNA LABELLE-HAHN

Harbor was about 1.7 miles away, we figured we needed a better option than having Ze crawl all the way.

We made her as comfortable as possible, got eight hundred milligrams of Motrin in her, and went about making a splint.

I used our InReach satellite text/GPS unit to send out an SOS to Rolf letting him know we had an emergency. Then I went off with a saw and began cutting lots of green willow and alder. Brianna got out the first aid kit and what tools we had. David took his backpack and hiked to Chippewa Harbor. Our plans had changed, and we knew we'd be in Chippewa Harbor that night camping. David would get us a campsite, but more importantly, if a boat was at the Chippewa Harbor dock, he would get them to use their marine radio to call the Park Service for help.

While I was cutting alder, a mother and son hiked down the trail and approached us. They greeted us and asked how I was doing (they didn't yet see Ze down the trail). I said, "Not so good, we have a broken ankle we're dealing with." They expressed sympathy and concern and continued hiking. They were headed to Chippewa Harbor too.

With a number of alder sticks and branches and some other wood pieces, Brianna fashioned a splint using duct tape. With the splint on and the Motrin having started doing its thing, we got Ze to her feet and attempted to walk her to Chippewa Harbor. She draped her arms over our shoulders and we took several steps with Ze having to hop awkwardly. The hopping made her foot bounce painfully. We quickly knew that walking her out this way wasn't going to work. We got Ze settled back on the ground as comfortably as possible.

I looked at Brianna and said, "We have to make a stretcher, and we'll carry her out." She gave me a blank look. I then said, "I'll cut longer, sturdier branches, and you have a lot of parachute cord—we'll make a stretcher with that and put a sleeping pad over the parachute cord." I could see Brianna's brain *click* into gear—she was an excellent medical device engineer professional with a strong MacGyver sense—and I knew we would have a good stretcher in a little while.

I went off to cut larger, sturdier lengths of alder. When I returned, Brianna had tied the parachute cord into some mystical macramé form that she slipped over the alder branches. Brianna made a few adjustments, we placed Ze's sleeping pad on the cord, and we had a working stretcher.

Ze rolled onto the stretcher and Brianna and I picked her up. We walked about thirty feet, realized we needed more people, set Ze down. It was too hot, the stretcher was heavy, and I was worried my grip on the wood would slip and we'd drop Ze. Ze again insisted she could crawl all the way out. It was not a serious option.

I checked the InReach to see if Rolf had responded to the emergency. He hadn't. I then sent a text to every address in the InReach which was only four or five addresses, but I figured at least one of them would see the message and respond soon.

David wasn't back, so I set off toward Chippewa Harbor. I met David shortly and he said there weren't any boats at the dock and the campground was empty except for the mother and son that had passed us on the trail. Mom and/or son would have to help us, I guess.

We returned to Chippewa Harbor and recruited Ronel and Ian to help. Ian pumped water for us and topped off our Nalgenes. He would stay in the campground—I told him if any boat came to the dock, he needed to let them know about the situation and tell them to contact the Park Service using their marine radio.

Ze was doing well. She was almost too cheerful given the circumstances. Ronel was a great addition—strong, steady, confident, and encouraging. Both had positive attitudes—just what our little group needed.

At that point, there were four of us to carry Ze, two on each side of the stretcher. It was still difficult, but conceivable. The toughest part was that most trails on Isle Royale were narrow, rocky, and uneven. Only two of us on one side could be on the trail at a time, while the other two needed to hike a step or two off the trail. We carefully carried Ze for around forty yards, found a spot in shade, and set her down to rest. Then we would repeat. We rotated around Ze so our hands and arms could have a break.

During one break, the InReach had a message from Sarah Hoy, one of the project's main scientists. She wondered if she should contact the Park Service. *Yes, please.* Then more carrying, rest, repeat. The sun remained very hot. At some point, the InReach had a message from Rolf saying that the Park Service was on the way. *Yes!*

We envisioned them coming in with professional gear and four big, strong men. Rescuers. Assuming we would no longer have to carry Ze, we took a longer break. It was a lot of work in the sun and heat.

In the end, we got Seth (a trained backcountry EMT) and Marshall. Not four big, strong men, but great just the same. Seth checked out Ze said she was doing okay, but needed attention. He liked the splint that Brianna had devised and left it for the time being. We continued to use our makeshift stretcher.

We continued carrying Ze with four of us while two could rest briefly. It was the same routine: Walk about forty yards, find shade, set down Ze, rest, drink water, and swap sides or rotate a rested person in. We made progress, and we knew we'd get Ze to Chippewa Harbor and onto a boat in the next hour or so.

Then two more Park Service people—Tony and Denise—arrived. Eight people could carry Ze. We made it so six of us were carrying her while two rested. The carrying was much easier, of course, but I especially

Carrying Ze and arriving at Chippewa Harbor.
SOURCE: BRIANNA LABELLE-HAHN

liked that with three of us on each side of Ze, it was less likely we would slip and drop her.

The 1.7 miles we carried Ze had three separate boardwalks with ten-inch-wide boards on little pylons over and through swamps. Carrying Ze with three of us on each side wasn't really possible on the boardwalks. This is where we let Ze crawl. She was good at crawling and did well on the boardwalks.

We finally got Ze to Chippewa Harbor Campground. Ian greeted us as well as two more Park Service people. There were two Park Service boats at the dock. We got Ze to the dock and onto the boat. Seth called in the situation and said they would be taking Ze across Lake Superior to

Parachute cord stretcher—just add a sleeping pad.

SOURCE: BRIANNA LABELLE-HAHN

Eagle Harbor where an ambulance would pick her up and take her to the hospital in Houghton.

Even though Rolf and Candy were probably listening to the Park Service radio traffic at Bangsund Cabin and knew what was going on, I InReach texted them the situation. We said tearful good-byes to Ze and she was gone. The Park Service boat zoomed out of the harbor and onto Lake Superior. She would be on the mainland in slightly under two hours and then in another forty-five minutes, she would be at the hospital in Houghton.

Weirdly, two days later, just after we had finished looking for dead moose for the day, we helped another hiker who had broken her ankle.

She'd broken it on trail and closer to the campground, so it was easier to get her out, but it was still a wash-rinse-repeat deal with a broken-ankle emergency. We got to see the same Park Service employees again (we were on a first name basis at that point).

Two days later, it continued to be excessively hot. During the afternoon, in the heat of the day, I was the only person in my group hiking. I came upon a hiker wobbling along the trail, not looking good. He was white as a ghost, had clammy-looking skin, and was sweating and panting a bit. I had figured I better talk to him. He—Paul—was experiencing heat exhaustion. We moved into the shade, got his backpack off, and I gave him my extra water with electrolytes. His wife—Barb—had gone ahead to leave her gear at the Lake Richie Campground where my group was. While at Lake Richie, Brianna, a member in my group, helped Barb. Brianna supplied Barb with water and electrolytes and hung Barb's gear in a tree to keep it away from wolves.

Barb planned to return to help Paul carry his gear. Instead, I carried Paul's backpack (and my daypack) about two miles to Lake Richie. Barb met us partway and helped Paul. When they got to the campground, they went into Lake Richie to cool off. Getting into the water helps immensely in the heat, lowering the core body temperature.

I'm pleased to say my group worked well together for each of these emergencies. We didn't get as many looking-for-dead-moose miles in as usual, but we handled the stressful situations well.

After Ze was on the boat heading to the mainland, Brianna said, "Is it okay if I had some Type 1 fun today?" Brianna enjoyed the challenge of making the stretcher and figuring out Ze's splint. I was fine with her having Type 1 fun since the stretcher and splint worked as designed.

Ze had surgery several days after leaving the island. After her surgery, she sent us an email:

> The emergency doctor and nurse were impressed with how stable my condition was, I explained the splint and ladder made of branches and wood pieces, and the heroic teammates that carried me out.
>
> I had trimalleolar fractures (three fractures in ankle). The surgery went well. . . . Now I am an iron woman with two titanium plates and about ten pins. A cast will be put on three weeks later after swelling from the

> surgery is gone. The surgeon decided not to fix the third fracture and will let it heal by itself. The second surgery will be removing the metal parts probably one year later.

Ze's surgery went well but she wouldn't be hiking anytime soon. Next year, for sure.

CHAPTER 34

WHAT I'VE LEARNED FROM BACKCOUNTRY OFF-TRAIL BACKPACKING

WHAT HAVE I LEARNED FROM BACKCOUNTRY OFF-TRAIL BACKPACKING? IT ISN'T car camping. It isn't parking at a trailhead and hiking a trail to an established campsite that appears on a map. It is being off trail in the wilderness, hiking with a full pack over forty pounds in all weather through swamps, and contending with big rocks, cliffs, ridgelines, creeks, deserts, and bugs. You don't know where your tent will be that evening. Any trouble you find is your trouble.

Off-trail backcountry backpacking can be confusing and uncertain. The volunteers in each group have to put their trust in a group leader they likely don't know a thing about. It can be uncertain, ambiguous, and uncomfortable. But most weeks end up being an absolute blast and are an experience to remember. Jump in with both feet, and enjoy it.

That isn't to say there aren't lessons to be learned. Here is what I've learned from off-trail hiking and leading groups.

Rule #1—The Basics

Don't be stupid, they're called *basics* for a reason—In life there are just some things you should *always* do. When in the backcountry and off trail:

- The first thing you do once you determine where you are camping for the night is GPS the location. You never know when you'll hike away from the camp and suddenly not know where you are or the direction you want to go.
- If you come across a good water source, get more water and top everything off.
- Be kind to your feet—Make sure you have good boots and dry socks, air your feet out periodically, and always have camp shoes. Trench feet, blisters, and detached toenails really happen and are—so I am told—painful.
- Always know where your hiking partners are—getting lost in the backcountry is super easy to do and can be scary.
- If—in the middle of the night—you have a "call of nature" and it is not raining, do it now. Do *not* wait. It might be raining later when you can no longer resist the call of nature. I know from experience that doing your business in a lightning-filled thunderstorm in a swamp at 3 a.m. is unpleasant. It is, however, memorable.
- Even if you don't think it is going to rain, put up a rain tarp in your camp.

Figure out, or know, what the basics are and do them.

Rule #2—Make Plans, Be Patient, Be Flexible

Blundering along off trail can be fun for a little while, but it really is best to have an idea of what you want to do.

- Just because you scoped out a six-mile hiking route in the morning doesn't mean you have to slavishly complete it. In fact, the odds are you won't complete it as planned.

- When hiking with a group, everyone has different speeds and sees different things, and there isn't any rush. The hike is not a timed event; it is a steady, slow progress. Take breaks often. You'll still get your miles in by the end of the week.
- Sometimes you know where you'll find what you're looking for, and sometimes you don't. Sometimes you won't even know what you are looking for, but you'll find it just the same. It is best to just keep your eyes open and be receptive.

Makes plans, but know they are something to deviate from.

Rule #3—Overpacking

Everyone knows not to overpack or not to overdo things, but it is an art that eludes most of us. For backpacking, a rule of thumb is you shouldn't carry more than 25 percent of your body weight; the ratio varies depending on who you ask. Regardless, it doesn't leave much margin for error on packing. Here's some thinking as to why the overpacking concept eludes hikers:

- Hubris—We all think we can carry more than we should. And we probably can, but over time, it can hurt in a number of ways and wear you down.
- A planning thing—Not realizing that backpacking and real life aren't the same thing and that different situations require different thinking. For instance, no, you don't necessarily need a new change of underwear for every day on the hike (sorry, Mom), let alone a second pair of pants. Just get used to the idea that you will marinate and stink by the end of the hike. So will everyone else.
- Planning again! There's a fine line between carrying too much and carrying what you need—talk to others in the group and split up communal gear. There's nothing like finding out you have three stoves *and* fuel.
- It's a long, tough hike, and some luxury is deserved. But that doesn't mean you can justify a milk frother, a bottle of wine, and so on. Even so, selective redundancy is good with items like stoves (just not

three of them), water filters and a GPS/compass/map. You might be able to get by with just one, but having two is better and you'll have a backup. Because—really—these items fall under the category of basics (rule #1).

Avoid over packing; know what you need, and don't take on too much more than that.

Rule #4—Lean Into Your Ignorance

Prepare, plan, and ask questions and acknowledge there are just some things you don't know.

- Test and know your gear before you go—Does it work, does it do what you expect, do you need it? There is nothing like carrying around extra weight if the gear doesn't perform, *and*, because in the backcountry, what you got is what you got, ain't no more.
- Asking for help is good—Someone else in the group will know the answer or can help you, and if they don't know or cannot help, then talking things over almost always means a solution can be found. Shared mistakes in the backcountry can actually be kind of fun.

Everyone is ignorant in their own special way. Get over it, accept it, and figure out how to work around your own ignorance. That's what teams are for.

Rule #5—You Are the Greatest Danger to Yourself

Don't be actively stupid, and you'll probably be okay and have a safe adventure.

- Wolves won't get you, and most every animal in the woods doesn't want a thing to do with you.
- If you're someplace with bears, they don't care about you either; they're interested in the food. Hang it or, better yet, vault it.

- Exceptions: Do *not* get between a mother and its young. Do *not* get involved when large animals are in a rut.
- Selfies with any animal is asking for it, especially Gray Jays (they're feisty). Do the *basics*, *be patient*, and *lean into your ignorance* and you'll be fine. Really.
- If all else fails, make sure you have a first aid kit.

If the saying "you're the greatest danger to yourself" is true, then why not go out and have an adventure? Get out there, now. Off-trail backcountry backpacking is about experience over things.

CHAPTER 35

WHAT THE MOOSEWATCH VOLUNTEERS CONTRIBUTE

WHO ARE THE VOLUNTEERS, THE CITIZEN SCIENTISTS, AND WHAT DO THEY accomplish? What is the result of all this hiking off trail week after week and year after year? And is it worth it?

> In 1988 Rolf began using . . . volunteers to help us with summer field work, and I was surprised by the success of this new approach. Not only were people willing to pay to help us, they were terrific at finding things. Many volunteers had such a great time that they returned year after year. I believe the personal relationships that have resulted are one of the most significant characteristics and benefits of this project.*

The remainder of this chapter is a business-like look at counts and averages, with rounding, for a single season of volunteer groups exploring the island. Averages are based upon informal counts (i.e., my personal

* Candy Peterson, Rolf's wife and a key member of the study since the early 1970s.

"data") from 2009 through 2022 along with averages from my groups over the years.

Eleven separate groups hike the island each season in four waves: three waves in the spring and one during the summer. The first two waves are generally on the western end of the island at Windigo and Washington Harbor, and the third and fourth waves are on the eastern end of the island landing at Rock Harbor and are based out of Bangsund Cabin for the week.

Every year, approximately fifty-six people volunteer to hike off trail to look for dead moose. This number varies from the mid-forties to the mid-seventies in recent years. Each group has from four to six people, including the group leader.

Groups will hike a total of thirty-five to forty miles per week with about twenty to twenty-five off-trail miles. There are sometimes more, sometimes fewer miles. This works out to about five or six miles per day, or, for all groups combined, about four hundred total miles per season, of which around 220 are off trail.

Volunteers find a dead moose or other "count" (e.g., wolf, fox, beaver) roughly every three miles of off-trail hiking, generating over seventy finds per season, or about 6.5 counts per group. Counts vary from year to year and group to group, of course. Some years are unusually successful in finding dead moose. During the 2022 Summer Study, volunteers and their leaders found 133 new records. It was primarily moose, but some fox and beaver were included in the count. The increased number of finds in 2022 were due to the newly introduced wolves from 2018–19 really settling in, reproducing, and becoming increasingly successful predators; the missed 2020 COVID season when nobody went to the island and there were dead moose just waiting to be found; and more volunteers and more groups than usual hiking that year.

There are a fair number of younger adults (age sixteen to twenty-five) and students that volunteer. Not surprisingly, these students are often natural science majors. There is a slight lull in volunteer counts aged from mid- to late twenties to about the mid-forties. Then counts pick back up as volunteers hit their mid-forties and older, with a number of volunteers in their seventies and even eighties.

Women represent about 35 to 40 percent of all volunteers, with, for some reason, a tendency for women to volunteer for the spring and the summer session. Counts for women and men are about equal before age thirty. Thereafter, the older a volunteer gets, the likelier it is a man.

Most volunteers come from the state of Michigan with healthy counts from Wisconsin, Minnesota, Iowa, and several other Midwestern states. People travel from other countries to volunteer, and not just from Canada. Volunteers have come from as far away as Great Britain, Costa Rica, Estonia, and Malaysia.

Isle Royale as a national park is known for having very few visitors, but those people who do visit the island are often repeat visitors. Similarly, Wolf-Moose volunteers are frequently repeat volunteers. In any given season, one-third to half of the volunteers will have been on one or more volunteer trips before. Some volunteers have gone out ten or more times. One volunteer has gone out over forty times.

The time and cost equivalent of so many volunteers is significant. The Wolf-Moose Project simply would not be able to hire this kind of help. Using eight-hour "work" days multiplied by the average number of fifty-six volunteers over an eight-day trip totals 3,584 hours for the benefit of the Wolf-Moose Project every year. If you were to assign a dollar value to this work—at $30 an hour—you would need at least $107,520 to hire this kind of help *for one year only.* The Wolf-Moose Project has been working with citizen-scientist volunteers since the late 1980s with the volunteers producing millions of dollars (or more) of benefit for the Wolf-Moose Project and science.

People volunteer for many reasons: because of their love of nature, because they've been to Isle Royale and learned about the opportunity to volunteer, for the challenge of hiking off trail, and because of the friends they've made volunteering previously. The volunteers are a dedicated bunch with many returning year after year for love of the island and love of the project. And as Candy Peterson said, "Personal relationships . . . are one of the most significant characteristics and benefits of this project."

CHAPTER 36

END OF THE WEEK ON ISLE ROYALE

EVERY BOOK HAS A LAST PAGE, EVERY MOVIE HAS A FINAL SCENE, AND EVERY week on Isle Royale comes to an end. Sometimes thankfully. On the western end of the island, the end of the week is at Windigo on Washington Harbor. This is the "quieter" side of the island, since more people enter and exit the island on the eastern end. A benefit of the western end of the island is you might score a shower at week's end in a Park Service building.

On the eastern end of Isle Royale, the end-of-the-week festivities occur at Bangsund Cabin. Two, three, and sometimes four groups descend on Bangsund at week's end, usually scattered over several hours through the afternoon, with everyone being dirty, tired, and hungry for salads. At least I know I am hungry for a salad.

None of this happens before 1 p.m., though. Rolf and Candy want time early in the day to get everything set up for the fifteen to twenty people soon to be overwhelming Bangsund Cabin. Bangsund has limited facilities, so they need to set up showers (i.e., tarps hung in the woods with a small piece of duckboard for a "floor"), acquire and prepare the food, and a myriad of other tasks to prepare for people and a lot of new dead moose parts.

Last-day headstand. The white speck of Rock Harbor lighthouse is visible in the distance.
SOURCE: SUE MORRISON

Before 1 p.m., most groups are hiking and gathering at Daisy Farm. Hiking at this time is done with considerably less urgency and more moseying—we've made it through the week and everything is great, even if it's raining. Moseying might shift into a bit of playfulness. Once, as we were hiking back to Daisy Farm on our last day, Sue Morrison saw a little sand bar in Rock Harbor and decided we needed to do headstands on it. So we did. I am on the far right, Sue is on the far left, and Sue's husband, Jeff, is in the middle.

Daisy Farm is the largest campground on the island and is immediately across about a quarter of a mile of open water from Moskey Basin where Bangsund Cabin is. Some groups might have been at Daisy Farm the evening before and done day hikes out of Daisy in the morning.

As late morning rolls around, there are frequently groups at Daisy Farm under the pavilion eating lunch. The pavilion is lovely because there are

End of the week 2022, approaching. Bangsund Cabin ready for food and to clean up.
SOURCE: ANNA BURKE

picnic tables and sitting on something other than the ground or a log is an option. If it is raining, the pavilion is protection. This last lunch usually consists of odds and ends of snack/lunch food from the week. Or a group might have had too much dinner food from the week and be making—and trying to share—real meals with everyone.

Often other Isle Royale visitors at Daisy Farm stop by and wonder what we've been doing for the week and why we have all the bones. They get impromptu mini lessons on the Wolf-Moose Project and the work the volunteers do. Several new volunteers have been introduced to the project at such mini lessons under the pavilion.

Eventually it is 1 p.m. and everyone wants to be at Bangsund. Rolf and Candy wait for us to communicate with them—in the last several years this means texting using InReach. We wait at the dock at Daisy and watch for Rolf to come out of Bangsund Cabin and get into a boat to go across Moskey Basin. Rolf then ferries us back and forth over the water, usually four or five at a time; this is about all the boat can handle with heavy packs, bones, and the volunteers.

As everyone arrives at Bangsund, we all have slightly different goals, and to an outside observer, it would probably look chaotic. But everyone has a logical set of chores or tasks they want to address. Candy will encourage us to take a shower; make us eat any baked goods she has pulled out of the oven (e.g., wild blueberry muffins are standard—they are delicious after a week of backpacking food); and personally check in with everyone to make sure they are healthy and had a good week. Rolf is like a kid in a candy store and mostly wants to look at the new bones and see our notes from the week. Not much else is on his agenda.

As it is not my first rodeo, I make sure the new volunteers know what they ought to be doing. For volunteers, the big items are:

- Take that shower that Candy keeps announcing loudly. Since there are only two "stalls" made of blue tarp, *and* because the hot water for cleaning is a large bucket of Lake Superior water heated on a gas stove, getting everyone cleaned up is a sequential, time-consuming task that takes most of the afternoon.
- Set up the tent for one last night at Bangsund before we catch a boat back to the mainland the following day. Bangsund has limited space for tents and we end up squeezing tents tightly next to each other.
- Throw away any garbage. All week we've been creating little bits of garbage and hauling it around following the tenets of Leave No Trace. Now is the time to dispose of the garbage; there is a single garbage can behind Bangsund Cabin.
- Return all borrowed project gear (e.g., water filters, GPS units, etc.).
- Relinquished all of the bones they were carrying. I have notes on who has what.

My first order of business is usually to grab a beer and sit in an actual chair. Cold items—including the beer—sometimes are in an antique Coca-Cola cooler behind Bangsund Cabin with large chunks of ice. Or they are in a milk crate sunk in Lake Superior just off the dock out front. The water temperature is usually around forty degrees—nice cold beer. Then I wander around and slowly get my group's bones lined up and make

Many tents in Bangsund Cabin's side yard on the last night. SOURCE: ALEC SMITH

sure I have my notes handy, because Rolf will be there shortly, if he isn't already there.

As time allows, I will go and find a spot to pitch my tent. Early on, I will still have some energy—knowing full well that I will lose initiative as the afternoon goes on—and also because of the limited space. I don't want to wait so long that there aren't good spots left. All the tents spring up in the "side yard" scattered around a fenced-in gravesite. In the tent photo, the picket fence (middle-right background) is the grave, purportedly somebody who was murdered after a drunken fight over one hundred years ago. The grave is perhaps twenty feet from the water, and me being skeptical and Isle Royale being exceedingly rocky, I have doubts there is enough soil at the "grave" to actually bury anybody.

Once everyone has showered, pitched their tent, and taken care of the project's gear, there are still a number of possible activities to do—assuming it isn't dinner time yet. And there is still energy to *do* something.

Rock Harbor Lighthouse. SOURCE: BRIANNA LABELLE-HAHN

From Bangsund Cabin, there is a little path carved through the cedars that goes to the Edisen Fishery and further on to the Rock Harbor Lighthouse. It is roughly a quarter-mile hike along the shoreline of Moskey Basin.

The Edisen Fishery is the most intact surviving example of a small, family-operated commercial fishery in continuous use on Isle Royale. It was in operation from 1910 to 1975 and represents a once-common lifestyle—a difficult lifestyle—on Lake Superior. It is a lovely little open-air museum.

Just past the Edisen Fishery is the Rock Harbor Lighthouse, now a museum. Built in the 1850s because of increased ship traffic around the island due to copper mining, it was the first lighthouse on the island and initially was in operation for several years up until the Civil War when it shut down until the 1870s. In the 1870s it went back into service, but was only in operation for several years when a newer, taller lighthouse—on Menagerie Island—was built, making the Rock Harbor Lighthouse superfluous. The Rock Harbor lighthouse hasn't been in operation since.

You can walk up the tower and, where the Fresnel Lens used to be, is a 360-degree viewing platform for some beautiful views. On the first floor of the lighthouse is the museum. It chronicles the many shipwrecks around the island (e.g., SS *Algoma*, SS *America*, SS *Congdon*, and many more). Most of these old wrecks from the mid-1800s to 1947 are intact and scuba divers can visit them. If you visit, make sure you close the door as you leave the museum (it is not staffed).

The lighthouse's fifty-foot tower has a decided, roughly two-degree tilt. Even so, the lighthouse is quite picturesque and right on the water near a dramatic passage from Lake Superior to the shelter of Moskey Basin.

Then—of course—there is the world's largest collection of antlered bull moose skulls. The Wolf-Moose Project brings bones back to Bangsund, and some are antlered skulls. Generally adult males do not die with their antlers on; moose, like deer, lose their antlers around December, a time when adult moose should still have plenty of fat reserves and be in relatively good health. As a result, antlered skulls are somewhat rare. And there are over one hundred of them at Bangsund—the world's largest collection.

The skulls are behind Bangsund Cabin arranged on boards and stumps. The skulls at the back are misshapen either due to old age when hormones

Peruke antlers. SOURCE: JEFFREY MORRISON

Salad—the stuff that backpacking dreams are made of. SOURCE: ANGELA JOHNSON

get out of whack, or the "peruke," which are weird antlers that were likely a genetic mishap.

The big event, though—after the showers and getting clean—is dinner. And that salad I wrote about earlier. Even though I try to make a point to have some fresh vegetables during the week (e.g., carrots, broccoli, sweet potatoes, etc.) it is a salad I start dreaming of midweek. The end-of-week Bangsund banquet always has a big salad with a random set of vegetables that come over on the boat. It doesn't matter what the vegetables are, they're in the salad and delicious.

Usually there will be a small bunch of volunteers inside Bangsund—or, if the weather is warm, on an outside picnic table—cleaning and prepping vegetables for the salad. The main dish is often lasagna—one meat and one vegetarian—that Candy has made ahead of time, along with at least two cakes. Dinner starts, a line forms, and people load up. Many people go

Singing "Hats Off" while Rolf, Candy, and the group leaders thank the volunteers.
SOURCE: ALEC SMITH

Bangsund Cabin sunset on the last night on the island. SOURCE: JACOB DEPPER

through the line two or more times to refill their plates. Sometimes people are in line a second time before everyone has gone through once.

And—of course—there is singing after dinner. Candy loves to have singing at week's end and many of the songs are re-purposed old girl scout songs; or so I am told, since I am not familiar with girl scout songs. The photo shows Rolf and Candy and three of us group leaders singing

Rolf, Candy, and a park ranger doing their Radio City Rockette routine as we leave the dock and island. SOURCE: JOHN STRYKER

something called "Hats Off" to thank the volunteers for their work on the island. I am the dashing-looking fellow with the wolf hat.

Eventually, the sun goes down we all go to bed in our sleeping bags, and sleep soundly. Bangsund has great sunsets.

The next day, we get on a boat and return to the mainland. One of the last sights we have of Isle Royale is of Rolf and Candy and anyone else who is handy giving us a song and dance: "Happy Trails" at minimum, and dance à la Radio City Rockettes as we leave the dock. It's been a good week on the island.

BIBLIOGRAPHY AND FURTHER READING

WEB RESOURCES

"Bangsund Fishery." National Park Service. https://www.nps.gov/articles/000/bangsund.htm.

"Edisen Fishery." National Park Service. https://www.nps.gov/articles/000/edisen.htm.

"Isle Royale." National Park Services. https://www.nps.gov/isro/index.htm.

"Moosewatch Expeditions." Wolves & Moose of Isle Royale. https://www.isleroyalewolf.org/volunteer-moosewatch.

The Wolf-Moose Foundation. http://www.wolfmoosefoundation.org.

The Wolf-Moose Project. https://www.isleroyalewolf.org/.

BOOKS AND FILMS

Allen, Durwand L. *Wolves of Minong: Their Vital Role in a Wild Community*. New York: Houghton Mifflin Harcourt, 1979.

Desort, George, dir. *Fortunate Wilderness: The Wolf and Moose Study of Isle Royale*. 2008.

Mech, L. David. *The Wolves of Isle Royale*. Washington, DC: US Printing Office, 1966.
Mech, L. David, and Greg Breining. *Wolf Island: Discovering the Secrets of a Mythic Animal*. Minneapolis: University of Minnesota Press, 2020.
Murie, Adolphe. *The Moose of Isle Royale*. Ann Arbor: University of Michigan Press, 1934.
Peterson, Carolyn C. *A View from the Wolf's Eye*. Houghton, MI: Isle Royale Natural History Museum, 2008.
Peterson, Rolf O. *The Wolves of Isle Royale: A Broken Balance*. Ann Arbor: University of Michigan Press, 2007.
Peterson, Rolf O. *Wolf Ecology and Prey Relationships on Isle Royale*. Washington, DC: National Park Service, 1977.
Vucetich, John A. *Restoring the Balance: What Wolves Tell Us about Our Relationship with Nature*. Baltimore: Johns Hopkins University Press, 2021.

ARTICLES

Adams, Jennifer R., Leah M. Vucetich, Philip W. Hedrick, Rolf O. Peterson, and John A. Vucetich. "Genomic Sweep and Potential Genetic Rescue during Limiting Environmental Conditions in an Isolated Wolf Population." *Proceedings of the Royal Society B* 278 (March 30, 2011): 3336–44. https://royalsocietypublishing.org/doi/10.1098/rspb.2011.0261.
Hoy, Sarah R., John A. Vucetich, Leah M. Vucetich, Mary Hindelang, Janet L. Huebner, Virginia B. Kraus, and Rolf O. Peterson. "Links between Three Chronic and Age-Related Diseases, Osteoarthritis, Periodontitis, and Osteoporosis, in a Wild Mammal (Moose) Population." *Osteoarthritis and Cartilage* 32, no. 3 (March 2024): 281–86.
"Moose Teeth Record Long Term Trends in Air Pollution." *Wolves and Moose of Isle Royale*. https://www.isleroyalewolf.org/moose-teeth.
Perkins, Cyndi. "What Studying Moose Bones for 65 Years Can Teach Us about Human Diseases." *Unscripted Research Blog*, Michigan Tech University, January 16, 2024. https://www.mtu.edu/unscripted/2024/01/what-studying-moose-bones-for-65-years-can-teach-us-about-human-diseases.html.
Vucetich, John A., P. M. Outridge, Rolf O. Peterson, Rune Eide, and Rolf Isrenn. "Mercury, Lead, and Lead Isotope Ratios in the Teeth of Moose (*Alces alces*) from Isle Royale, U.S. Upper Midwest, from 1952 to 2002." *Journal for Environmental Monitoring* 11, no. 7 (2009): 1352–59.